HISTORY'S MOST
DANGEROUS JOBS
MINERS

HISTORY'S MOST
DANGEROUS JOBS

MINERS

ANTHONY BURTON

Front cover image: Carn Brea mine, at the 274 fathom level. A triangular staging composed of ladders and planks from which a hand-labour team are drilling their holes (1890s). Photo: J.C. Burrow FRPS.

Images facing title page. Top: Shovelling coal at the coalface in 1895. (National Mining Museum, Scotland); *bottom*: A boy harnessed to a cart, as seen in the 1842 Parliamentary Report on Children in Mines. (Author's collection)

First published 2013

The History Press
The Mill, Brimscombe Port
Stroud, Gloucestershire, GL5 2QG
www.thehistorypress.co.uk

© Anthony Burton, 2013

British Library Cataloguing in Publication Data.
A catalogue record for this book is available from the British Library.

ISBN 978 0 7524 8478 5

Typesetting and origination by The History Press
Printed in Great Britain

CONTENTS

INTRODUCTION

I have been writing about different aspects of industrial history for some forty years but the world of mining has always held a special fascination for me. As a boy I once went potholing with a friend, but found the experience of scrambling around in dark, dripping caves distinctly unpleasant and unrewarding. Yet I have cheerfully crawled around equally damp and dripping tunnels if they happened to be part of an old mine complex, never quite knowing what I would find. I remember a particular Derbyshire lead mine where, to my astonishment, I came across remnants of a very old underground railway, with rails made not of iron but out of wood. I was also fascinated by working mines, and the more I visited the more respect I developed for the men who worked in these places, doing a job that I felt I would never be able to manage. But, of course, if I had been born in a different place and at a different time, I would have had little choice in the matter. I recall one old miner telling me the story of his early years. Like many mining families, his parents were not keen for their boy to go down the pit, and he managed to get a job in a local factory. He thought the work was going well, until he was called into the manager's office and handed his papers. When he was asked what he had done to get the sack, he was told that his father was down the mine and that's where he must go now. The factory owner had an arrangement with the mine owner not to take on any boys from colliers' families. Down the mine he went and there he stayed.

I was struck by the injustice of this and other stories, and the more I read of mining history, the more I sympathised with the miners who struggled for generation after generation in two great battles – to survive in a hostile environment and to attain a decent living for exceptional work. My book *The Miners* was published in 1974, and I was interviewed on radio, television and in the press. One interviewer remarked that she thought the book was very interesting but she did feel it was a bit one-sided. I could only plead guilty. For me the story of mining, and especially coal mining, is one of oppressors and the oppressed, and there aren't two sides to that equation. I have not tried to distort the facts to fit some preconceived political or social attitude. There is no need: the facts speak for themselves. So I am afraid some people might find the present book one-sided as well, but I hope that readers will feel as I do that that is the nature of the story not a biased opinion forced upon it.

When I wrote the first book I wanted it to reflect what I had felt meeting and talking to miners, and hoped that it would accurately reflect their lives and that of their forebears. I have had a few decent reviews over the years, and a few not so good, but nothing has given me greater pleasure than a handwritten poem sent to me by a face worker, which he simply called 'The Miners – for Anthony Burton'. I am quoting it here not just out of a sense of personal pride, though I am immensely proud to have received it, but because it contains a truth about the story I shall be telling here, a truth about continuity and hardship from one who has lived the life:

> Miners' praises are rarely sung
> By someone away from the mines
> Once in a while there comes a man
> Who realises the meaning of hard times
> He researches, he studies to find out the truth
> Though sometimes he may be misled,
> For the truth for which he is looking
> May be buried with those who are dead
> But stories are carried down the years
> From grandfather to father to grandson
> He puts them all down to make up a book
> About miners
> As a miner I'll say well done.

I hope that many will find this often tragic story to be one worth reading, but above all I hope that all who are working or who have worked in mines will think I have done them justice.

Anthony Burton

CHAPTER ONE

BEGINNINGS

Today, mining is no longer at the centre of our economic life. The change has been dramatic. The metal mines have virtually all closed, and even the once mighty coal industry is reduced to a rump. Where over a million men worked down the pits a century ago, the most recent statistics show just over 3,000 working in deep mines. Once productive pits have either been abandoned or kept open as museums. Visitors can go down and see for themselves where their forebears worked, but they will never be able to get more than a hint of what a working colliery was really like. Even in modern times, when mechanisation has taken over much of the back-breaking work of extracting the coal, the life of a miner is unique. At the coalface itself, lit only by the lights on the miners' helmets, the automatic cutter carves its way through the coal, delivering great chunks on to a conveyor. The air seems to be filled with black rain and the visitor, unused to such places, soon finds that walking with a permanent stoop produces aches in muscles they never knew they had. The miners will be happy to point out that things were once a lot worse. One man told me of joining his father working in a seam that was just 2ft 10in high. His father, he said, was a big man and he wondered how he could ever fit into such a narrow space: 'He just chucked his belly up and climbed in after it.' I visited several collieries while working on an earlier book in the 1970s, and I remarked to one miner that I'd still been blowing black dust out of my nose days after a previous visit. 'Ay, lad,' he said, 'but it's what you don't blow out that'll kill you.' It was a potent reminder that working in a mine means a life fraught with dangers. Most people are aware that this is the case, but it is easier to overlook the inconveniences; the things that the rest of us take for granted. Once the men go down at the start of a shift, they stay down until the shift is over: no nipping out for a coffee or going out to lunch. The only food and drink they have will be what they have brought with them. Sanitation is non-existent: all you can do is dig a hole and bury the evidence. In this book I shall be looking at the long history of mining in Britain and the men who worked underground – and the coal miner is, in fact, one of the more recent arrivals on the mining scene. Mining can be described as man's oldest industrial activity, dating back literally thousands of years. Inevitably, there have been immense changes over such a long period of time, but certain essentials remain. It is an activity that takes the miners into an alien, hostile and always dangerous environment. It is no coincidence that in many fables and religions, the world beneath our feet is the world of the dead: we ascend to the heavens in the skies if we are virtuous; otherwise we descend under the earth into hell. Given the dread in which this underworld is held, it is remarkable to find that we have been exploiting it in Britain since the New Stone Age, almost 5,000 years ago.

Britain at that time was heavily forested. If people were to change their way of life from hunter-gatherers to agriculturalists, they had to clear the trees and till the ground to raise crops and to feed animals. They needed tools, sharp tools, and the best available material was flint. It can be shaped by striking with a hard stone, a process known as knapping. Small flakes can be adapted for making arrowheads and larger pieces can be shaped and polished to create axe heads; the better the quality of the flint, the better and sharper the tools that can be made from it. Nodules of flint can be found on the surface in chalk lands, but the people of the New Stone Age must have discovered at some time that the best flint could be found deeper underground. They became miners.

Grimes Graves in Norfolk is an area of rather scrubby chalk land, covered by what appears to be a bewildering maze of humps and hollows. From the air it looks like a First World War battlefield, scarred with massive craters. In fact, these are collapsed flint mine shafts and there are estimated to be over 300 of them spread over around 14 hectares. The pits were sunk over a period of approximately 500 years, the earliest dating back to 2,300 BC. One of the pits has now been opened up to the public, but I was fortunate enough to have the opportunity to go down another pit and explore the complex of passages that lead away from the bottom.

The deepest shafts are about 15m and sinking them involved an immense amount of hard work. It has been estimated that each pit probably took at least twenty men four months to dig. They seemed to have worked steadily down to the bedrock, the layer of high-quality flint, and from there they began working outwards in galleries. They did not waste their energies making passageways to walk down: they made them no bigger than absolutely necessary, and today it is an uncomfortable crawl along the narrow openings, in places scarcely big enough to squeeze a body through. When they came to a really productive band of flint, they excavated larger chambers, though still hardly big enough to kneel in to attack the flint. There was a limit to how far it was practical to move from the main shaft. The flint and spoil had to be passed out down the narrow passageways and they must have decided that it was a better use of manpower to dig a new shaft and start again once the journey became too long and unmanageable.

The system I visited ended in a chamber, with a thick band of black flint at floor level. There was also a hole, like a window, in part of the wall. The miners had been digging away when they broke into an older working, so it was pointless to continue into an area from which the flint had already been removed. Today you can shine a light through, and see a working face that has been undisturbed for around 4,000 years, left just as it was when the last Neolithic miner crawled away and abandoned the workings. You get a glimpse of one of the tools these early miners used: a deer antler propped between the floor and roof. It would have been used as a pickaxe to lever the flint nodules out of the chalk. Elsewhere archaeologists have found animal shoulder blades, used as shovels, and little bowls moulded out of chalk that once held animal fat or oil, which would have been burned to provide light. The conditions must have been very difficult. When I turned my lamp off, the darkness was as absolute as if a velvet hood had been pulled over my head, and the little flame of the miners' primitive lamps could hardly have been more than a flicker in the blackness. It is still an eerie place to visit and it is not surprising that a small chalk figurine was found, which one assumes was an offering or a talisman to protect the miners.

The flint mines must always have been insecure places, as indeed all mines always have been. When William Crago first went down a copper mine to start work at the age of 9 in the 1870s, he went with his father and before the first descent his father prayed for their safety. He may have prayed to a different god from that evoked by his Neolithic forebear, but the impulse was the same. Those working underground need all the protection they can get. The similarities do not end there. Young William had only the light of a candle to guide him in the mine; not that much more effective than a bowl of burning oil or fat. He too had to make his way from the pit bottom to the face, though his descent was far deeper and the underground passage far longer. But though separated by many centuries, the two generations of miners were essentially doing the same thing: digging down to the valuable material, whether ore or flint, and doing whatever they could to ensure their safety along the way.

When I emerged from my visit to the flint mine, covered in chalk where I had been forced to wriggle on my belly through the narrower openings, I had a real admiration for those workers of long ago. Popular mythology has the people of the New Stone Age as little better than dumb animals, but the labyrinth of tunnels had been created with care and a real understanding of how much chalk could be removed in safety, and what had to be left to avoid collapse. These were true miners and this was a genuine industry. There was far more flint removed than could ever have been needed locally, and archaeological research has identified axes made from flint from Grimes Graves spread right across southern England. Grimes Graves was part of an extensive trading network. It was here and in other similar mines that Britain's mining history began.

Somewhere around 2000 BC men began experimenting with materials that were an improvement on stone: they began making implements of bronze. This is an alloy of copper, which in its later development usually meant the addition of tin. This was a major step forward in technology as the metals only occur in the form of ores, which have to be smelted (heated to a high temperature in a furnace of some sort, to release the pure metal). Tin and copper can both be found in south-west England and it is well known that Phoenicians came to Cornwall to trade in tin. At this stage in development the tin could mostly be found in surface deposits, washed out of the surrounding granite. Mining was not greatly advanced in those regions, but in other parts of Britain things were very different.

Archaeologists have discovered extensive remains of copper mining at Great Orme's Head in North Wales, the oldest parts of which date back to the Bronze Age.[1] The ore would have been easy to spot, where the brightly coloured veins appeared at the surface, and at first extraction would only have involved following the vein down through the rock. But soon they would have discovered that there were other rich veins, branching out in different directions. To work these would involve creating passageways through the rock. This was a far more difficult task than simply hacking away the comparatively soft chalk of sites such as Grimes Graves. Stone tools have since been recovered from the site, but these alone would not have been sufficient. They used a technique known as fire setting. This is very much what the name suggests: a fire is lit against a rock face and allowed to burn fiercely. After a suitable period of time, cold water is dashed against the red-hot rock, which splinters. Recent research has suggested that the effect is more from the heat than any contraction caused by sudden cooling. It must have been a very unpleasant operation in confined spaces deep underground, with passageways filled first with acrid smoke and then clouds of steam. As all this happened long before the arrival of written records, we have no idea how many, or indeed if any, miners were overcome by fumes. The surviving passageways carved out of the rock are impressive testimonies to the success that was achieved using the most primitive tools.

Around 500 BC there was another leap forward in technology as we see iron begin to replace bronze. A much harder metal, iron making resulted in a huge improvement in the manufacture of tools and weapons. Once again everything depended on extracting the ore, and iron mines joined the swelling ranks of underground workings. Clearwell Caves, in the Forest of Dean, were certainly worked during the Iron Age as many of the names still in use centuries later were Celtic in origin. For example, the Free Miners of Dean, who own traditional rights in the forest, still refer to their holdings as 'gales', from the Celtic *cael*, meaning 'to have'. However, although the end product was different, the working methods had scarcely changed since the preceding Bronze Age. Things began to change again with the coming of the Romans.

The Romans developed mines for a wide range of minerals, sometimes following on from workings begun by the Celts. Pliny the Elder, who died in AD 79, noted that 'In Britain lead is found near the surface of the earth in such abundance that a law is made to limit the quantity that shall be gotten'.[2]

The Romans undoubtedly increased production in the mines of Derbyshire and the Mendips. Eighty pigs of lead have survived that can almost certainly be dated to the Roman period, each weighing an average of 184lb. In the eighteenth century, when the Derbyshire lead industry was at its peak, pigs weighed 176lb: not much of a change at the end of 1,600 years of development. Lead was important to the Romans, who used it as a waterproof material for everything from lining baths to manufacturing water supply pipes. The Latin name is *plumbum*, which is why the people who install water systems are still known as plumbers. But the Romans also developed mines to extract more exotic materials, and the systems they used brought a new level of sophistication to mining practice.

> Pliny the Elder indicated that lead was being mined near the surface in Britain before the Roman invasion.

Just outside the hamlet of Pumsaint in Mid Wales is the site of the Roman gold mines of Dolaucothi.[3] Local people were employed in the mines, though it is likely that they were little more than slave labour. They did, however, enjoy a life under Roman rules and regulations, which included a requirement to provide a bathhouse for the miners; a luxury that only reappeared in every mine in Britain in the twentieth century. The work, however, differed very significantly from anything that had gone before. All the mining sites discussed so far have been based on little more than the strong arms of the miners to extract the ore, or the simple task of lighting a fire. The Romans made extensive use of hydraulic power, and that meant supplying large quantities of water to the site. The trouble was that the mines were high up a hillside and the only source of water, the nearby rivers Cothi and Annell, were down in the valley. Nevertheless, there was one technique that the Romans had mastered over the years: they knew how to build aqueducts. They needed to be able to use efficient surveying techniques to discover points on the rivers that were higher than the mine site before they could begin work, and it turned out that the longest aqueduct carried water 7 miles from the Cothi. The watercourses can still be traced, and they are a tribute both to the surveying and engineering skills of their makers. The gradients are gentle, and near the actual site parts have been cut right through solid rock. Large reservoirs were then constructed above the potential mines, and the main work of locating and extracting the gold could begin.

The first stage of development was to find where the ore outcrops reached the surface, and this involved removing the topsoil and vegetation. This was where water played its part, in a technique known as 'hushing'. Water from the tanks was directed down the hillside in a powerful jet, washing away the surface material. The gullies, scoured out using this technique, can still be seen on the hillside. Once the veins of ore had been located and mining had begun, the stored water was used for washing the crushed ore, to separate the valuable gold from the waste, rather as later generations of miners panned for gold. Water, however, is not always the miners' friend. As they go deeper underground, miners will eventually meet water, which has to be drained away. One solution is to make adits (channels cut to allow water to drain away); a long and complex process that involves carving passages through the rock to emerge on the hillside. The Romans, however, had another new technology to offer. They were the first to develop the Vitruvian waterwheel, one that turns on a horizontal axis and named after the writer who first described it, Vitruvius. Remains of such a wheel were found about 50m below the surface. It would have acted rather like an overshot wheel in reverse: the latter was turned by water being fed along a trough to fall in buckets set round the rim of the wheel, this in turn scooped up the water and carried it up to fall into a trough to be drained away. It is likely that a series of wheels had to be used to raise the water to the surface.

The actual mining techniques had scarcely changed since the Bronze Age, still relying on fire setting as the main method of breaking up the rock. Some of the workings show smooth sides that indicate this method was used, but in other places there are still marks left by pickaxes as the workers hacked their way through the rock. Today the site is in the care of the National Trust, and parts of the workings have been opened up to visitors. It is a complex site as the mines were reworked in the nineteenth century, but it is still possible to recognise the Roman features. Perhaps the greatest tribute one can pay to the Romans is to note that their handiwork is almost indistinguishable from that of their Victorian successors. Almost the only way to tell later workings from the earlier is to look for signs of drilled holes that were packed with explosives.

Dolaucothi miners were able to use iron tools, instead of earlier stone or bronze, and could install waterwheels to keep the mine drained. Very few new advances were made in mining technology for many centuries after the Roman occupation of Britain ended: a Roman time-travelling miner would have felt quite at home in a medieval mine. Mining continued to be a vital part of the British economy, and over time many different types of mine were opened up, from graphite mines in the Lake District to slate mines in North Wales, but one type was to prove the most important of them all, and it was one of the last to be developed: the coal mine.

In the rest of this book, the emphasis will be on coal miners and the tin and copper miners of Cornwall and Devon. There are inevitable technological similarities in the two methods, but the way in which the work was organised was completely different, and these differences had profound effects on the lives of the miners. But whatever type of mine a man went down, he could be sure that the conditions would be harsh and dangers ever present. At the beginning of this chapter an analogy was drawn between the physical underworld and the underworld of mythology. It was an analogy not lost on those who worked in the industry. Pit deputies in coal mines played much the same role as foremen in other industries. The official history of the Midland Area branch of their union begins with this little poem, which may be humorous, but it is a bitter humour:

> A Deputy stood at the Pearly Gates,
> His face was worn and old,
> He meekly asked the Man of Fate
> Admission to the Fold.
>
> 'What have you done,' St. Peter asked,
> 'To seek admission here?'
> 'I ran a district in the pit
> For many, many a year.'
>
> The gate swung open sharply,
> As Peter touched the bell,
> 'Come in,' he said, 'and take a harp
> YOU'VE HAD ENOUGH OF HELL.'[4]

CHAPTER TWO

T'OWD MAN

'**T**'owd man' is a name given to miners of long ago, sometimes applied to spirits and goblins said to live down mines, and more specifically to a carving of a miner on a plaque in the church at Wirksworth in Derbyshire. It has never been accurately dated, but has been variously described as being Saxon or early medieval. One thing does seem quite certain: this is the oldest representation of a British miner. He is very simply dressed with a long tunic worn over hose and his only head protection appears to be a skullcap. Over his shoulder he has his pickaxe and he is carrying a basket to collect the ore. A very similar figure can be seen on the emblem of the Free Miners of the Forest of Dean, which has a slightly later date. The finest set of illustrations of early mining techniques can be found in a sixteenth-century treatise on metal mining.[1] The costume has changed very little, though many of the workers have hooded tunics. If they look vaguely familiar it is because Walt Disney used them as a pattern for his seven mining dwarves. These sources give us an image of how miners must have looked when they went off to work, over a period of many centuries.

In the early medieval period, mining still meant mining for metal ores and, unlike the industry of later years, much of the work was still left to individuals or small partnerships. In Derbyshire most of the lead-mining area had belonged to the Crown since Saxon times, and was known as the King's Field of High Peak. Indeed, the laws governing mining in the area date from 'a time whereof the memory of man runneth not to the contrary'.[2] These laws allowed any man to dig for ore wherever he liked in the King's Field, with the exception of

The emblem for the Free Miners of Dean shows a man in fourteenth-century costume, with a pick in his hand, his hod over his back and a candle at the end of a clay stick held in his mouth. It is very similar to the carving of T'owd Man in Derbyshire.

under churches, in churchyards, in the highway or in orchards, though he was required to pay his dues to the king or his lessees. There are numerous stories of landowners planting a single apple tree in a field in the hope that it would be considered an orchard and keep prospectors out. The ancient laws were challenged in the thirteenth century, but an inquest held at Ashbourne upheld the rights of the miners. The techniques of mining were still fairly crude and the nearest thing to mechanisation was the method used to get the ore from underground to the surface. A bucket or kibble was raised and lowered by means of a simple hand-operated winch or scowse. A miner who had found a vein that he wanted to work would put up a scowse and, when he had acquired a suitable amount of ore to fill a 'dish', he took it to the Barmaster in one of the special Barmote Courts. It is a good indication of just how long this system had been in place that the name 'Barmaster' is an adaptation of the Saxon *bergmeister*, a superintendent of mines. The court held a standard dish. One of these, which has a capacity of 472 cubic inches, has survived, and carries a suitably detailed, official inscription on the side:

> This dishe was made the iij day of October the iij year of the Reigne of Kyng Henry the viij before George Erle of Shrewesbury Steward of ye Kyngs most Honourable Household and also Steward of all the honour of Tutbury by the assent and consent as well of all the Mynours as of all the Brenners within and adjoining the Lordshyp of Wyrksyworth percell of the said Honour. The Dishe to Reamyne In the Moote Hall at Wyrksyworth hanging by a cheyne so as the Merchauntes or mynours may have resorte to ye same at all tymes to make the trew mesur of the same.[3]

The dish remained the standard measure and was used to determine how much levy the miner would be paying to the Crown, as well as providing a standard to be met to justify a claim. Once the Barmaster was satisfied that the legal requirement had been met, he allotted a length of one 'meer' on each side of the 'possession scowse', a measurement that varied from one area to another but was between 29 and 32yd. A third meer was granted to the lessee. Meers could be worked by a miner on his own behalf, but they could also be owned by wealthy individuals or institutions: in Saxon England important mines near Wirksworth were owned by the Abbess of Repton.

Once an individual had been awarded his meer, each end was marked by a scowse and he could set to work. He might come across different types of vein. Rakes are narrow fissures, running more or less vertically down from the surface and possibly extending over a considerable length. The ore would be removed by following the vein down, usually until the water level was reached. The landscape of Derbyshire is still scarred by the remains of these rakes, which appear like rocky canyons running through the countryside, often for hundreds of yards. Pipe veins are created in fissures in the limestone, formed by water. For the prospecting miner they represented a gamble. Some that appeared thin at the surface never developed, but some opened out into wide caverns. A good pipe represented rich pickings, because the ore was only loosely held in soft material and was easily extracted. The disadvantage was that the pipes were invariably very wet, so working conditions were extremely uncomfortable. The least productive veins were known as flat work: thin layers between the limestone bedding planes.

Individuals could work their own mines simply because so little capital outlay was required: all one needed was a pickaxe, hammers and wedges. Once the ore was brought up, it had to be dressed to remove the valuable material from the waste. This work usually went to women, who broke away the unwanted material by beating it off with a 6lb hammer. The final separation was not much more sophisticated. Once most of the waste had been knocked off, the remaining material was put in a jigger, a perforated box, which was shaken in a tub of water. The heavy lead-bearing material sank to the bottom. Then the ore was ready to be sent to the smelters.

Special mining privileges were not limited to Derbyshire. In medieval England, the land was divided into hundreds, areas from which the Lord of the Manor was required to provide a hundred fighting men if the king required them. The hundred of St Briavels in the Forest of Dean sent men to help Edward I take Berwick-upon-Tweed from the Scots. It was heavily fortified, but the men of St Briavels had just the skills required for the capture of the town: they literally undermined it. They had always mined in the forest, but now their rights were officially recognised. Anyone who had been born in the hundred and had worked a mine for a year and a day could be registered as a Free Miner by the king's representative, the gaveller. These rights not only still exist, but are still exercised. The first Free Miners worked for iron, but their successors are colliers.

When one reads of medieval miners working on their own it seems unlikely, because we have an image in our minds of a modern colliery employing hundreds of men. But in the Forest of Dean you can still find collieries worked by a handful of men or even individuals. Robin Morgan is a true Free Miner. As he said, he used to bunk off school to go down a mine with his brothers, and he has always had a mine of his own throughout his adult life. Today he works Hopewell Colliery and, although it is not identical to one of the old medieval workings, there are enough similarities to give a good idea of how many early mines were developed. He works a drift mine, one which is tunnelled directly into a hillside, rather than sunk as a vertical shaft. Not that the tunnel is level: it slopes quite steeply down to the coalface. The seam is quite narrow and difficult to work, and no more than 3ft high. He uses a modern cutter to remove the coal, but still has to shovel it out by hand, lying on his side on clammy clay. He loads the coal into trucks or 'drams' on a simple railed track. They are then hauled up by cable using an engine from an old

Morris car, which wasn't even new when he bought it thirty-four years ago.
Robin Morgan's working life is like that of earlier miners: it consists of a good
deal of hard, physical labour and calls for ingenuity and adaptability.

The other principal metal-mining area of medieval Britain was to be
found in the south-west, in modern-day Cornwall and Devon. It is a region
principally known for tin in this period, but it was also an important source of
a precious metal: silver. It seems that silver was first discovered in two sites, at
Combe Martin in Devon and at Bere Alston in the Tamar valley, at the end of
the thirteenth century. Silver normally occurs in association with lead and the
miners would have found some ores that glittered more brightly than the rest.
News of the discoveries reached the king, Edward I, who promptly used the
royal prerogative and claimed it all for himself. Records show that between the
'twelfth daie of August and the last of October' 1294, 370lb of silver ore was
sent to the king from Martinstowe, modern Maristow. In 1297, the silver from
the same source was valued at £4,046[4] (roughly the equivalent of £2 million
at today's values). Another indication of the very high value of the silver ore
can be gauged by the fact that much larger quantities of lead removed in the
same year were worth only £360.

The lead was being extracted from quite shallow workings, but the silver
lay deeper underground and was more difficult to work. At this time there
was very little expertise in deep mining in this part of the world, but the king
was not about to allow that to get in the way of filling his exchequer. Expert
miners were needed so he issued a decree that 340 men from Derbyshire
would be sent to work in the Combe Martin mines. They were not given a
choice: refusal to go meant imprisonment. So they had to leave their families
and make the 250-mile journey on foot from Derbyshire. In time, the mines
at Bere Ferrers on the River Tavy proved so lucrative that it was no longer
necessary to force men to work them. It was possible to pay high wages and
attract miners not just from Derbyshire but also from continental Europe.
These men brought new techniques to the work, driving adits, known locally
as avidots, through the ridge that separates the Tavy from the Tamar. In the
early days of shallow pits, work was limited by the seasons: winter meant short
days and bad weather meant all work could come to a halt. The new, deep
workings meant that seasons and weather no longer had any effect, and silver
production doubled.

By the fifteenth century, the mines of the south-west were employing
more than a thousand men. Bere Ferrers, which had been a hamlet before
the work started, gained borough status and a market, where a curious law
forbade miners from 'loitering under the pretence of buying meat'. Clearly
the authorities felt that their proper place was underground, making more
money for the king and his representatives. The miners did, however, enjoy
special privileges as royal employees, including exemption from some tolls and
taxes, and if they were considered guilty of a crime they were brought before

a warden instead of the usual court. Punishments were generally less severe in the warden's court, though a pit was dug at Bere Ferrers to act as a prison for the 'contrairient and rebellious'.

By the fifteenth century, the region had developed into an important mining area, in which the ore was treated as well as being removed from the ground. It was smelted on site, at first in kilns like the familiar limekilns still to be found in the area; later in special furnaces, where extra heat was supplied by forcing air through the fire using bellows, powered by a waterwheel. After smelting the silver was cast into ingots, which gave us a name still in use for our currency. The monetary pound was set as the value of 1lb weight of silver. It has changed a little since those days, as at the time of writing that weight of silver would fetch over £3,000. Real industrialisation was coming to the area, though the actual mining techniques had scarcely changed. What had changed was the money the men could make. Fourpence a day might not sound like much, but it put the silver miners right at the top of the earning tree for manual workers, and was probably the first time that the special skills of miners had been recognised in terms of their pay.

Although the extraction of tin in Cornwall and Devon goes back into prehistory, it was not at first produced by mining. A mythology has grown up surrounding the early history of tin. In Elizabethan times it was accepted that people had come from the Near East to trade, but there was a great deal of confusion about who they actually were. We know them as Phoenicians, but in the Middle Ages people generally either called them Saracens or Jews. There were stories of brass ornaments in Solomon's temple made out of tin and the tradition lives on in Penzance with the market in Market Jew Street. There are stories of St Paul arriving to preach in Cornwall and buying tin, and that Joseph of Arimathea was a tin-worker. Real records, however, only take us back to the twelfth century AD.

Tin first attracted royal interest when Richard I set off to campaign in Europe, leaving Hubert, the Archbishop of Canterbury, to look after his financial affairs. Hubert decided that revenue could be gained from tin, using a system very similar to that established in the lead-mining district of Derbyshire. He set up a code of laws governing the trade, established standard weights and measures, and divided the region into stanneries governed by stannery courts. A Charter of the Stanneries was written in 1202, confirming the tinners' privileges 'of digging tin, and turfs for smelting it, at all times, freely and peaceably and without hindrance of any man, everywhere in the moors and in the fees of bishops, abbots and counts'.[5]

The system of marking out claims was simple:

When a mine (or stream work) is found in any such place, the first discoverer aymeth how farre it is likely to extend, and then at the four corners of his limited proportion, diggeth up three turfes and the like (if he list) on the sides, which they term Bounding, and within that compasse, every other man is restrained from searching.[6]

The result was a patchwork of small holdings, the names of which give an idea of how very personal they were: some are quite jovial – Goodmorrow Neighbour; others optimistic – North Goodluck; and some simply bewildering – Pitpaddy and Park Buggens. Few had any great value. My wife's Cornish ancestors held tin bounds, and one of them, Melchisedech Rogers, left a will when he died in 1732 in which his bounds were valued at £4, his tools at 5s. To put that in context, his feather bed was valued at £2. It was not a system that encouraged large investment in expensive operations, but in the early years this was not necessary. Tin was washed out of veins in the granite and could be obtained by streaming. This is a technique similar to panning for gold. Pebbles from a stream are swirled with water, separating out the heavy tin from the lighter stones. There is a survivor from those days at the Blue Hills Tin Streams in Cornwall. One can still pick up ore from the beach; you can soon appreciate the difference between these pebbles and the rest. A bucket of tin is remarkably heavy. Tin streaming continued throughout the medieval period, but actual mining was comparatively rare in the south-west, and any pits that were dug were shallow and small.

While minerals were being extracted from the ground all over Britain, coal was still virtually ignored until the early thirteenth century. The likeliest explanation is that, in an age when few houses had proper fireplaces or anything more complicated than a hole in the roof instead of a chimney, the smoke would have been intolerable. In any case, there was little need to look for an alternative fuel when wood and peat were readily available. But if domestic use was limited, there were others who would appreciate the fact that burning coal could produce high temperatures: just the thing for blacksmiths' forges and limekilns. In fact, coal was comparatively easily obtained where it outcropped at the surface along the shores of the Tyne and the Firth of Forth, and could often be picked up from the beach. It was known as sea coal, and as early as 1228 a street leading down to the Thames in London, near what is now Ludgate Circus, was called Sacoles Lane. It is the first indication that coal was being shipped from the north to the capital.

As the century progressed, references to coal began to multiply: in 1256 the monks of Newminster Abbey in Northumberland had been given the right to collect sea coal. After that, more and more rights were acquired by a variety of different bodies and coal mining became common in Northumberland, Durham and on the Firth of Forth. It soon spread throughout the coalfields of Britain. By the reign of Edward III, the coal trade was sufficiently important for regulations to be needed. It was in this period that official measures were introduced, based on the keel. The keel was a barge used to transport coal and was defined as carrying twenty chaldrons, with each chaldron holding 2,000lb, so the total load would have been roughly 18 tons. In Newcastle any merchant who was not a freeman of Newcastle had to pay 2*d* a keel to the Crown.[7]

At first, coal mines were comparatively shallow, never venturing below the level at which natural drainage was possible. Some were drift mines, but others were worked from a vertical shaft. The simplest versions consisted of a shaft dug down to the coal level, from where the men worked out from the bottom of the shaft in all directions, creating a large cavity. The shape gave them the name of 'bell pits'. The only mechanical devices used were the simple winches to haul the coal out of the pit. The similarities to Grimes Graves are obvious. Here, too, there was a limit to how far away from the shaft men could work without the risk of the roof collapsing, and when that point was reached they simply abandoned the pit and started another. Some mines were worked from galleries dug horizontally from the foot of the shaft.

Drainage was chiefly by adit. The system was satisfactory for shallow mines and was continued far longer in the lead mines of Derbyshire. By the eighteenth century, adits, or as they were known locally, soughs, were driven over great distances. One of the most imposing was the Hillcar Sough. It was started in 1766 to drain the mines at Alport-by-Youlgreave and ran for about 4 miles, eventually emptying into the River Derwent. This was an immense undertaking that took years to complete and then ended up costing £50,000.

Driving such adits demanded a great deal from the miners:

> Few operations can be conceived more unpleasant and dangerous to the workmen, than the execution of these adits, especially when, as sometimes the case, they are barely wide enough to allow the sinker to creep along. The dangers which are created by blasting the solid rocks with gunpowder in such confined spaces, will be easily conceived.[8]

The same technology was used in the Cornish mines, where the average size of an adit was 6ft by 2½ft. This meant that only one man at a time could work at the face, so advancing an adit was slow work. It was recorded that a major adit could take up to thirty years to complete. It was also dangerous:

> Whenever they are apprehensive of coming towards the house of water, as the miners term it, they bore a hole with an iron rod towards the water about a fathom or two or so many feet further than they have broke with the pick-axe. As they work on, they still keep the hole with the barrier before them that they may have timely notice of the bursting forth of the water, and so give it vent or passage. Yet notwithstanding all this care and prudence, they are often lost by the sudden eruption of the water. In some places, especially where a new Adit is brought home to an old mine, they have unexpectedly holed to the house of water before they thought themselves near it, and have instantly perished.[9]

Pure water was not the only obstacle that might be met. One group of miners driving an adit found their way crossed by a forgotten sewer. There was nothing to do but clear it out. The man who was given the job eventually emerged with his clothes filthy and distinctly aromatic, and he had no way of cleaning himself up or changing until he got home. On the way he stopped off to get a wad of consoling baccy, but was promptly shown the door by the shopkeeper, complaining that he was likely to turn the meat bad just by breathing near it.

In the fifteenth century, the collieries of Tyne and Wear were beginning to push down below the water level. In 1492–93, the records of the Bishop of Durham have an entry for a payment for 'two great iron chains for the ordnance of the mine at Wickham, for drawing coal and water out of the coal pit there, by my lord's command'. Mining was becoming big business, and the days of individuals working their own claims were numbered.

CHAPTER THREE

THE DEEP PITS

The opening up of deep mines brought profound changes to the nature of mining and the lives of miners. The most obvious change was the need to bring in expensive equipment to drain out water, equipment that the old independent miners could never afford. Newcastle's first historian, writing in 1649, gave one of the earliest descriptions of the new system:

> Some south gentlemen have upon great hope of benefit come into this country to hazard their monies in coal pits. Master Beaumont, a gentleman of great ingenuity and rare parts, adventured into our mines with his thirty thousand pounds; who brought with him many rare engines not known then in these parts; as the art to bore with iron rods to try the deepness and thickness of the coal; rare engines to draw water out of the pits; wagons with one horse to carry coal down from the pits to the staithes at the river, &c.[1]

Sadly, it seems he never managed to make a profit 'and rode home upon his light horse'.

Various devices were used to raise water, of which the most common were the rag-and-chain pumps and Egyptian wheels. The former consisted of bunches of rags fitted at regular intervals to a chain, which could move up and down in a tube, though it was later made more efficient by replacing the rags with metal plates. These acted like a succession of pistons, working in cylinders. The Egyptian wheel was similar to a more modern bucket dredger. One of the more imposing Egyptian wheels was used by Sir George Bruce to reopen a drowned-out colliery at Culross in Scotland. It consisted of an endless chain fitted with thirty-six buckets and was driven by three horses. The buckets dipped into a well at the bottom of the pit and emptied into a trough at the surface. The colliery extended for a mile under the Firth of Forth and was considered as one of the wonders of the age. King James I came to see it and was taken down the shaft on the shore of the Firth. It was essential to have a second shaft to keep the mine ventilated, and as the other end of the workings was beneath the water, it had to be sunk from an artificial island. The king was not aware of this, and when he emerged and found himself surrounded by water, he assumed the worst and promptly started calling out 'Treason!'. It seems to have taken some time to pacify him.

These seventeenth-century workings were still comparatively shallow to today's, but as pumping methods became ever more efficient, so mines went deeper and deeper. This made life hard for the miners. At the beginning of the eighteenth century Daniel Defoe travelled all over Britain, taking careful note of everything he saw, including local industries.[2] In Derbyshire he stopped off to visit a lead mine:

We went … to a valley on the side of a rising hill, where there were several grooves, so they call the mouth of the shaft or pit by which they go down into a lead mine: and as we were standing still to look at one of them, admiring how small they were, and scarce believing a poor man that shew'd it us, when he told us, that they went down those narrow pits or holes to so great a depth in the earth; I say, while we were wondering, and scarce believing the fact, we were agreeably surprised with seeing a hand and then an arm, and quickly after a head thrust up out of the very groove we were looking at. It was the more surprising as not we only, but not the man that we were talking to, knew any thing of it, or expected it.

Immediately we rode closer up to the place, where we see the poor wretch working and heaving himself up gradually, as we thought, with difficulty; but when he shewed us that it was by setting his feet upon pieces of wood fixt cross the angles of the groove like a ladder, we found that the difficulty was not much; and if the groove had been larger they could not either go up or down so easily, or with so much safety, for that now their elbow resting on those pieces as well as their feet, they went up and down with great ease and safety … Besides his basket of tools, he brought up with him about three-quarters of a hundred weight of oar, which we wondered at, for the man had no small load to bring, considering the manner of his coming up; and this indeed made him come heaving and struggling up, as I said at first, as if he had great difficulty to get out.

Thanks to the ability to work ever-deeper pits, coal production in Britain rose from nearly 3 million tons a year in 1700 to over 10 million tons by the end of the century.

A simple transport system was used underground in the sixteenth century. Trucks with wooden wheels were pushed along planks and were kept on track by means of a pin at the bottom of the truck running in a groove.

Defoe went on to describe how this man, who clearly worked hard in difficult circumstances, actually lived:

> There was a large hollow cave, which the poor people by two curtains hang'd across, had parted into three rooms. On one side was the chimney, and the man, or perhaps his father, being miners, had found means to work a shaft or funnel through the rock to carry the smoke out at the top … The habitation was poor, 'tis true, but things within did not look so like misery as I expected. Every thing was clean and neat, tho' mean and ordinary: there were shelves with earthen ware, some pewter and brass. There was, which I observed in particular, a whole flitch or side of bacon hanging up by the chimney, and by it a good piece of another. There was a sow and pigs running about at the door, and a little lean cow feeding upon a green place, just before the door, and the little enclosed piece of ground I mentioned, was growing with good barley; it being then near harvest.

In nineteenth-century Cornwall, descent down the mines was made easier by the use of proper ladders, but was no less arduous than it had been a century earlier in Derbyshire. William Crago, who first went to work down a copper mine there in the middle of the nineteenth century at the age of 9, gave a graphic account of what it was like to descend a deep pit for the first time. Before setting off he collected candles. Gas was not a problem in these mines, so the men simply stuck candles on their hats to light their way. He joined his father, who had to carry a great deal of equipment down, and even the 9-year-old boy was expected to do his bit: 'I had on my right arm about five pounds of black powder carried in a copper can, on my left, a coil of fuse to be used with the powder for blasting purposes, in my pockets gad, used for splitting rocks, and each of us a fair sized potato pasty for our dinners.' They waited until all the other men were well on their way before beginning their own descent. They went down 20ft, and there they paused to light their candles:

> We at last stepped into the footway, Father first and I following him, very carefully we descended the first 480 feet. It was almost like climbing down the side of a house and as we slowly went along, ever and anon came Father's warning voice 'Hold tight your hands, my son'.
>   As I have before said, the first portion of our downward course was like climbing down the side of a house. The portion of the shaft we were now about to descend was termed the underlie Section … When the ladders are upright the strain of climbing comes almost entirely on ones arms whereas if the ladders are on the slope the feet have to bear a much larger portion of the weight of the climber's body, and so I found on resuming my descent, it was very much easier to climb down this portion than I had hither found it, but with that advantage I found there was attached to it a great disadvantage. You will readily imagine that

in most mines there is a great quantity of water and as a rule the Cornish mines are very wet indeed.

It is not only the water that is found while sinking the shafts and driving the levels, but surface water which finds its way down gives a great deal of trouble to the miner engaged in sinking operations and very much inconvenience when climbing up or down.

It is not pleasant to have a small stream of water dropping just on the back of your neck and running down your back and legs into your boots and then have to work in that state for 7 or 8 hours, but such was the fate of most of those who worked in that portion of our mine.

We continued what seemed to me our interminable journey down into blackness, my legs were aching, my back was stiff and sore, and my hands were getting benumbed with the continual clutching of the ladder rungs and it was quite necessary that Father's warning shouts be given 'Hold tight your hands'.

After about two hours climbing (I was soon able to do it as quickly as most boys) we reached 1,600 feet landing stage, and I for one felt extremely thankful to step out of the ladder onto what appeared a much safer place, but in order to enter the subway we had to cross the shaft on what appeared to be a very narrow plank and great caution is necessary in so doing, however, this was safely accomplished and we entered the mouth of the subway.

All this had to be achieved before this young boy started his seven- to eight-hour shift, and then at the end of the day he had to climb all the way back up again.

The long climbs up and down the ladders every day seem like desperately hard work, but for some women in Scottish mines things were even tougher. A contemporary observer of the Scottish industry, Robert Bald, described their work: Many of these young women had families. They were up at dawn and had to carry their babies, bundled up in shawls, to a baby-minder before making their way down the mine. The real work started at the coalface, where men loaded coal into wickerwork baskets, which were hoisted on to the women's backs. Bald measured these loads and found them a literally staggering 170lb. The women then carried them 150yd up a steep slope to the foot of the shaft. They next climbed ladders for 177ft to the surface, and once at the top they still had to go another 20yd to reach the coal store, where they could empty their loads. The women would make this journey as many as twenty-four times in a shift, which could last anything from eight to ten hours. A strong woman would carry almost 2 tons of coal a day, during which time she would have climbed almost 3,000ft with the load on her back and carried it for another 3 miles. It is hardly surprising that Bald reported 'it is no uncommon thing to see them, when ascending the pit, weeping most bitterly, from the excessive severity of the labour'.[3] For this the women received the princely sum of 8d per day. Even that is not quite the end of the story. The practice had been banned in the Glasgow area at the end of the eighteenth century, but Bald found it was still widespread in other Scottish mines at the beginning of the next century.

The lot of the women in Scotland was unusually harsh. There were two distinct classes at work. The married women and daughters of working pitmen were generally able to stay within the family group, which gave them some protection from the worst excesses. The rest were known as 'fremit' or framed bearers. They were attached to no particular miner and were sent wherever the overseer decided. If the man they were sent to work for proved particularly ill tempered, they had no choice but to accept whatever curses or violence he inflicted on them. Many of these unfortunate young women were burdened with the heaviest loads.

The working day did not end for these women when they left the mine. They had babies to collect and feed, and they returned home to a cold, empty house with no fire. They were still in their pit clothes, always filthy with coal dust and often soaking wet. They did what they could to provide some sort of comfort for their families, yet it was all but impossible to maintain any sort of standard of decent living. The result was inevitable: disease was rife and infant mortality reached appalling levels. Bald noted many Scottish mining communities where the infant death rate exceeded the birth rate. Bald was disgusted by what he saw. He described meeting one of the women who was:

> groaning under an excessive weight of coals, trembling in every nerve, and almost unable to keep her knees from sinking under her. On coming up, she said in a most plaintive and melancholy voice: 'O Sir, this is sore, sore work. I wish to God that the first woman who tried to bear coals had broke her back, and none would have tried it again.'

There was little the Scottish miners could do to ease their burdens. Much of Scotland was impoverished following the Jacobite rebellion of 1745 and work was hard to find. The miners' houses were virtually all owned by the mine, which effectively prevented them from leaving to find other work. According to Bald it was not only the houses that were owned by the employers: the miners themselves were no better than serfs:

> They belonged to the estate or colliery where they were born and continued to work and from it neither they nor their children could remove; so much so, that when an estate with a colliery came to be sold, the colliers and their families formed part of the inventory or livestock, and were valued as such; in short, they were bought and sold as slaves.

Over the years many improvements were made in overcoming the difficulties of getting the coal from the face to the surface. The horse gin was a simple but effective device. It consisted of a winding drum, with a horizontal wooden beam mounted on the vertical axis of the drum. A horse was harnessed to the beam and walked round a circular track to turn the winding drum; depending on which way it walked, the rope wrapped round the drum either unwound to lower a kibble down the shaft or was wound on to raise it. The other problem that remained to be solved was moving the coal to the foot of the shaft. The big breakthrough came in 1776 when the mining engineer John Curr developed a system using trucks, or corves, running on a plateway of iron rails. He published a book in 1797 describing the system, in which he claimed that twelve corves could be hauled as a train using a horse gin, and that it would be possible to shift as much as 250 tons a day down a 250yd roadway. It was eventually realised that it might actually be more efficient to take a small horse or pony down the pit and let it pull the train along. This was such a benefit to the mining community that the system was even praised in verse, which was written in the dialect of Tyneside:

> God bless the man wi' peace and plenty
> That first invented metal plates,
> Draw out his years te five times twenty
> Then slide him through the heavenly gates.
>
> For if the human frame to spare
> Frae toil an' pain ayont conceivin'
> Has aught te de wi' getting' there,
> Aw think he mun gan' strite to heaven.[4]

Robert Bald, who had fulminated against the treatment of women in Scottish pits, was among those who praised the new system and the benefits it brought. He found that miners' wives were able to stay at home and make the house into the comfortable place it had never been before. The working men and boys came back to a fire and a meal. The house became a place in which the whole family took pride. He noticed that there was a new love of impressive furnishings: 'A chest of mahogany drawers, and an eight-day clock, with a mahogany case, are the great objects of their ambition.' The clock, in particular, was seen as an emblem of success in life, and when it was finally acquired it was put on display and all the neighbours were invited in to admire it. Other commentators, such as John Holland, noted the same fascination with furnishings in the collieries of north-east England:

It would be impossible for a stranger to pass in front of the lowly dwellings, three or four hundred in number, adjacent to Jarrow Colliery, for example, without being struck by the succession of carved mahogany bed-posts, and tall chests-of-drawers, as well as chairs of the same costly material, which are presented at almost every door.[5]

It is not too surprising that the colliers showed off their expensive furniture: it was all they had. The Jarrow houses were all grouped round the pithead, were of poor quality, with little or no sanitation, and they were, like their Scottish counterparts, owned by the mine. Anyone who left the colliery to work elsewhere was removed. These were poor people, just about adequately housed, who spent their working days in the darkness of the pit. Once above ground they wanted a bit of ostentation and bravura in their lives – and deserved it. Holland went on to describe the colourful dress of the men on their days off:

> In their dress, the pitmen, singularly enough, often affect to be gaudy, or rather they did so formerly, being fond of clothes of flaring colours. Their holiday waistcoats, called by them *posey jackets*, were frequently of very curious patterns, displaying flowers of various dyes; their stockings mostly of blue, purple, pink or mixed colours. A great part of them used to have their hair very long, which on work-days was either tied in a queue, or rolled up in curls; but when they drest in their best attire, it was commonly spread over their shoulders. Some of them wore two or three narrow ribbons round their hats, placed at equal distances in which it was customary with them to insert two or more bunches of primroses or other flowers.

It seems distinctly odd to think of miners dressing up like eighteenth-century hippies, but for them it was an affirmation that their lives did not have to be drab all the time, nor entirely taken up with back-breaking labour. This desire to express themselves was seen at its most colourful at special events, as this description from *The Newcastle Journal* makes clear:

> October 14, 1754. William Weatherburn, pitman, belonging to Heaton, was married at All Saints' Church, in Newcastle, to Elizabeth Oswald of Gallowgate. At the celebration of this marriage there was the greatest concourse of people ever known on a like occasion. There were five or six thousand at the church and in the churchyard. The bride and bridegroom having invited their friends in the country, a great number attended them at church; and being mostly mounted double, or a man and a woman on a horse, made a very grotesque appearance in their parade through the streets. The women and the horses were literally covered with ribbons.

The grim picture painted by Bald suddenly appears not quite so grim as some advances at least made life easier for the families, and particularly for the women. There were even reports of leisure activities in the eighteenth century, such as keeping racing pigeons, which are associated with mining communities right up to the present day. However, improvements were patchy; many mines still continued in the old ways and for far too many women and children there was little improvement; in some cases life became even harder. The actual job of winning the coal or the ore had scarcely changed since medieval times, and was still the preserve of men.

There were two main methods of working coal. In the north of England they mainly used a method known variously as pillar and board or pillar and stall. In this method, the seam is worked in the stall, but enough coal is left standing as pillars to support the roof. The size of the pillars depended very much on how much weight they had to bear, and the deeper the mine the bigger the pillars needed to be. In the early eighteenth century, a board was generally never more than 3yd wide and the actual work of breaking out the coal from the face was done by a single miner, the hewer. It was calculated that by using this method, less than half the coal in the seam was actually removed. This was obviously wasteful, but attempts to win more coal could lead to the collapse of a board or even the closure of the whole mine. The alternative was known as longwall working. In this system, the whole vein was removed, but the spaces left behind were filled with rubble or supported by props, usually in the early days made up of piles of stone. Headways led from the vein back to the shaft. There were two possible variations on this method: longwall advancing and longwall retreating. In the former, the wall is advanced from the bottom of the shaft. In the latter, the headings are dug away from the shaft to the boundary of the coal, and the coal between the headings is then worked back towards the shaft. The advantage of retreating is that as the face advances, the props in the worked section can be removed, allowing the roof to collapse behind the working face. The disadvantage is that while the headings are being dug, no coal is being dug. Whichever method was used, pillar and stall or longwall, the work was arduous, but just how difficult depended on the thickness of the seam and the hardness of the coal. As Bald described at the workings in Scotland:

It appears that in working the Scotch coal, which is very strong in the wall, it requires such constant exertion and twisting of the body, that unless a person has been habituated to it from his earliest years, he cannot submit to the operation. For instance, it is a common practice for a collier, when making a horizontal cut in that part of the coal which is upon a level with his feet, to sit down and place the right shoulder upon the inside of his right knee; in this posture he will work long, and with good effect. At other times, he works sitting with his body half inclined to the one side, or stretched out his whole length, in seams of coal not thirty inches thick.

The reader might like to try that sitting position, which feels more like something you might attempt in a yoga class than a posture for yielding a pickaxe for hours on end.

Even as late as the beginning of the eighteenth century, the dangers of coal mining were little understood. An anonymous author described the dangers that awaited the inexperienced miner:

> He may lose his life by Styth, which is a sort of bad foul Air, or fume, exhaling out of some Minerals … and here an Ignorant Man is Cheated of his life insencibly, as also by his ignorance may be burnt to Death by the Surfeit, which is another dangerous sort of bad Air, but of a fiery Nature like lightning, which blasts and tears all before it, if it takes hold of the Candle, which an experienced Labourer will discover and extinguish.[6]

Styth, also known as choke damp, is carbon dioxide, which suffocates, and Surfeit, or firedamp, is mainly methane. What is notable about this description is the fact that the miners were working using the naked flames of candles. There were attempts to detect the methane: a candle flame would start to burn blue; there was a distinctive smell; and some miners took dogs down the pit, as they are more sensitive to smells than humans. The only remedial work was to try to improve ventilation by the use of upshaft and downshaft and a series of 'stops' intended to ensure the 'fresh' air reached all parts of the workings. This was the system recommended by the anonymous author previously quoted, but as he admits it was far from infallible:

> But now I would give you an Account, that for the safety of the Miners, we must be careful of guiding the Air under Ground, lest we bring a Styth, or a Blast, by the Sulphur or Surfeit, upon the poor Men, as of late it did, not far from *New-Castle*, I think it was but in *October* 1705, that I was told by one who was acquainted with, and see some of the Dead buried, and had been at the Pits after the Blast, that there was above Thirty Persons Young and Old slain by a Blast, perhaps in less than Minutes time. How it came to pass he could not give me an Account.

He added the grisly detail that one boy was blown clear out of the shaft from the bottom of the mine, with half of his head missing. At this date explosions were comparatively rare, but new developments in technology would only increase the dangers.

The copper and tin mines were safer than the coal mines in that they were never gaseous, though the threat of flooding was ever present. Working methods were very different, in that the veins or lodes of ore were seldom horizontal like bands of coal, but could slope very steeply. The men who followed them had to work from a series of stages, boring the rock with hand drills, before filling the holes with black powder for blasting and removing the ore. The mines also suffered more from fluctuations in demand and prices than the coal districts, and the fortunes of the miners themselves changed with them. As a result, the workers in the early years of the eighteenth century suffered greatly from the threat of starvation, a problem made far worse by what was widely seen as the greed of the corn-factors, who sent the grain out of the county to fetch higher prices, leaving none for the locals. The result was a series of riots throughout the period. In 1729, the local tinners broke open stores and took away the corn. There was an immediate response from the authorities. Ringleaders were named, arrested and executed, and their bodies hung in chains on St Austell Downs as a warning to others. In spite of the severity of the response, the starving miners had little option and similar scenes were repeated time and again throughout the county, always with the same result. Some mining families ended up with much-needed food in their bellies; others were left mourning for members of the family shot by soldiers or hung by the courts. The dangers of living in a mining community were not all to be found underground.

CHAPTER FOUR

BIRTH OF THE STEAM AGE

As mines went ever deeper, so the problems of drainage increased. Relays of waterwheels were sometimes used to raise the water in stages, but there was a depth beyond which they could not work efficiently or indeed at all. Something new was needed; that something turned out to be steam. The first practical steam pump to be used in Britain's mines was the invention of a military engineer, Captain Thomas Savery. He filled a vessel with steam and then condensed it, creating a vacuum. Water was next sucked up to fill the vessel. Steam was then again passed into the vessel, forcing the water out into a second vessel at a higher level. He called this device 'The Miner's Friend', patented it, and by 1702 he was advertising his invention:

> Captain Savery's engines which raise Water by the force of Fire in any reasonable quantities and to any height, being now brought into perfection and ready for publick use; These are to give notice to all Proprietors of Mines and Collieries which are incumbered with Water, that they may be furnished with Engines to drain the same, at his Workhouse in Salisbury Court, London, against the Old Playhouse, where it may be seen working on Wednesdays and Saturdays in every week from 3 to 6 in the afternoon, when they may be satisfied of the performance thereof, with less expense than any other force of Horse or Hands, and less subject to repair.[1]

The invention was not quite the success that Savery envisaged in the mining industry. In spite of the claims made in the advert, there was a limit to the height at which it could raise water and the invention had to be installed at the bottom of the mine. This was not a good idea in a coal mine, where lighting fires to boil water is generally not encouraged: many working miners must have looked on their new 'Friend' with considerable distrust.

At almost the same time, another inventor down in Devon was also thinking about using steam power to drain mines. Thomas Newcomen was a merchant working in Dartmouth and among his most successful lines of business was supplying the local miners with iron tools. It was also said that he had trained as an engineer, though no documents recording an apprenticeship have ever been found. Whether he had formal qualifications or not, he soon showed that he had an inventive mind and understood basic engineering principles. A Swedish engineer, Mårten Triewald, came to England in 1716 and confirmed that Newcomen worked without any knowledge of Savery's invention and that he had an assistant, a plumber called Calley. He hoped to patent his invention, but Savery already had a patent that covered all engines using 'The Impellant Force of Fire'. Newcomen's engine might have been entirely different but it still fell foul of that catch-all phrase. There was nothing to be done by Newcomen other than to reach agreement with Savery and to pay him royalties. It was a good bargain for Savery, for Newcomen's engine was far superior to his own machine.

Mines were no longer reliant on the old-fashioned rag-and-chain or Egyptian wheel pumps. The newer, more efficient pumps used stout vertical timbers dropping down the shaft, ending in a piston working in a cylinder. As the pump rod moved up and down, so the water could be lifted. The down stroke was no problem, gravity saw to that, what was needed was a force to raise the pump rods up again, and this was where Newcomen used steam. The pump rods were suspended from one end of an overhead beam, pivoted at its centre. At the other end of the beam he hung a piston, fitted into a cylinder. He built a copper boiler, like an overgrown kettle, and used it to raise steam that was fed into the cylinder, forcing out the air. He then dowsed the cylinder with cold water, creating a vacuum. Air pressure now acted on the top of the piston, forcing it down. As it dropped, so the rods at the other end of the beam were raised. Once the pressure had equalised, the whole cycle could be repeated. The giant beam bobbed slowly up and down, and the pump rods rose and fell. The water could be raised in a series of lifts, all worked by the same engine.

The first recorded Newcomen engine was installed at a colliery at Dudley Castle in 1712. There is a story that the first engines employed boys to work the valves in the correct sequence, but one of these ingenious lads got tired of the repetitive task and using 'catches and strings' arranged for the valves to be worked by the engine itself. It is probably just one of the many homely fables that surround all inventions and no more reliable than the rest. What is absolutely certain is that the steam pumping engine was a great success, and the nodding giants were soon to be seen at collieries and mines throughout the land. They were massive affairs that required their own special engine houses, in which one wall – the bob wall – was much thicker than the rest as it had to act as the main support of the overhead beam. The only way to get more work out of an engine was to increase the size of the cylinder, and by the middle of the eighteenth century engines were being built with cylinders as much as 6ft in diameter. The Newcomen engine was a vast improvement over all the older methods, but it had one disadvantage: it required a lot of coal to keep the boiler going. This was not a problem at collieries where the one commodity they had in plenty was coal, but was a major difficulty in the metal mines of the south-west, where the nearest source of fuel was South Wales. Many of the men in charge of the mines, the mine captains, were inventive engineers in their own right and were more than happy to try their hands at improvements, often with some success. Among them was a Cornish mine captain called Richard Trevithick, whose son was to make an even greater advance in technology when he built the world's first steam locomotive to run on rails. That, however, was in the nineteenth century; the eighteenth-century engineers were still desperately trying to save on fuel costs and wondering how it was to be done. The answer was discovered at the opposite end of the country, not in a mine but in a university workshop in Glasgow.

James Watt was the university instrument maker and he was sent a model of a Newcomen engine that refused to work. He recognised where the problem of excessive fuel use arose. At every stroke of the engine, the cylinder was alternately cooled and heated, and he began to think of ways in which the cylinder could be kept permanently hot. His answer was the separate condenser: instead of condensing the steam in the cylinder he would do so in a separate vessel. But he could not keep the cylinder permanently hot if the top was open to the atmosphere, and if he closed the top then he could not use air pressure to move the piston. However, he could, he realised, use steam under pressure: the combination of steam pressure on one side of the piston and a vacuum on the other would do the trick. He had turned the Newcomen atmospheric engine into a genuine steam engine. He formed a partnership with the Birmingham manufacturer Matthew Boulton, and together they created a company that was to become world famous.

The Cornish miners were desperate for the new engines. The agent for Wheal Virgin – 'Wheal' is the normal name for mine in Cornwall – wrote: 'The existing engines at full power and the mine will be forced to close unless some answer is found … the engines last month used 300 Weys of coal … which sweeps away all the profit.'[2] The Boulton and Watt engine was just what was needed. The efficiency of an engine at that time was measured by its 'duty', defined as the numbers of pounds of water raised 1ft high by the consumption of one bushel of coal. The original Newcomen engine was rated at 4.5 million; later, when improved by the great engineer John Smeaton, it reached 12.5 million, but the first Boulton and Watt engine did 22 million and that figure

James Boswell visited the factory where Boulton and Watt manufactured steam engines. In his *Life of Johnson* he wrote: 'I shall never forget Mr. Boulton's expression to me: "I sell here, Sir, what all the world desires to have – POWER".'

was later improved. Boulton and Watt charged a premium for the use of their engines, the equivalent to one-third of the saving between using their engine and the old Newcomen engine. The Cornish, it appeared, could not lose. In the years between 1777 and the end of the century, Boulton and Watt supplied more than fifty engines to Cornwall.

There was, however, one snag. James Watt had a patent that was so all-embracing that it effectively stopped anyone else developing any aspect of the steam engine without his approval – and he never gave his approval. This did not stop the young Turks among the Cornishmen from trying. The Cornish engineers and mine captains had always been independent and ingenious. Where Richard Trevithick Senior had made improvements to the Newcomen engine, his son Richard was soon working on his own ideas, and he was far from alone. Others, such as members of the Hornblower family and Edward Bull, also worked on new designs. To Boulton and Watt, these men were simply pirates and they attempted to use the force of the law to stop them. One of the problems they faced was serving papers on the miscreants. One bailiff arrived at a mine intending to serve a notice, but he was recognised. The miners grabbed him, tied a rope round his middle, suspended him over the shaft and politely enquired if he was sure he wanted to serve the papers. He felt that, on the whole, he would rather not. The steam engine may have brought huge benefits to the Cornish industry, but that did not stop the local men's resentment of anything that stood in the way of their own right to think for themselves. The Cornish engineers did manage to make some advances in design that didn't infringe the patent: Trevithick produced a much improved boiler that eventually became known as the Cornish boiler, for example. But real advances had to wait until the Watt patent expired in 1800.

In time, James Watt himself would make further improvements to the steam engine, including the replacement of the flexible chain of the old engines that suspended the piston by a fixed linkage. As the end of the beam moves through the arc of a circle, he had to find a device that allowed that to happen while still keeping the piston rod moving up and down in a straight line. He introduced his parallel linkage, a hypnotically fascinating device of sliding rods that he always regarded as his most ingenious invention. Under the old system, the engine was always single-acting – you can't push a beam up using a flexible chain. Now, however, there could be two power strokes. This meant that the engine could now be adapted to other uses: it could become a winding engine, taking over the work that had once relied on the plodding horse walking round the gin, and in metal mines it could be used to work stamps to break up the ore. It was a godsend to the Cornish, and the evidence remains in the engine houses that are still such a feature of the landscape. Mines could now be made more profitable and worked at ever greater depths. This presented few problems in Cornwall, other than to the miners who had even more ladders to go up and down, but brought colliers fresh dangers.

As mines became more complex, so the dangers of firedamp became more difficult to control through ventilation alone. The most common method of ventilation was to set a furnace at the foot of the upshaft: the rising hot air would pull cooler air down the second shaft and through the workings. This could involve air travelling immense distances: in the Walker Colliery in north-east England, the air travelled through 30 miles of galleries before reaching the upshaft, by which time it must have been unbearably hot and scarcely breathable. Even in the most modern collieries, the difference in temperature between the air at the end of its long journey and at the beginning is considerable; in earlier times it would have been far worse. Temperatures underground could be so high that men were forced to strip naked in order to work in anything approximating to comfort. As mines became more complex, the air had to be forced round ever-more circuitous routes to ensure all parts of the pit were properly ventilated. A system was developed using trapdoors to direct the flow of air, but it only worked so long as the doors were shut, except for the brief periods when men and corves (tubs used to move the coal) were being moved. The job of opening and closing the doors went to one of the saddest creatures in mining history – the trapper boy. These young boys, who might be no more than 6 years old, regularly worked up to thirteen hours a day, crouching on their own in the pitch dark, with nothing to do but open and close the doors when required. This was typically the first job any boy got when he first went down the pit.

It was generally understood by the end of the eighteenth century that firedamp was such a real danger that naked lights should not be used in a coal mine. However, light was needed, and so several devices were tried. One of the most popular was the steel-mill, which produced sparks that were thought to be safe; they were not. The most bizarre idea that was tried was to use the glow caused by the luminescence from putrefying fish. It seems incredible that anyone could try to work with such a feeble glow, let alone cope with the inevitable stench, but no one came up with any practical alternatives, nor did anyone in authority seem greatly concerned.

When gunpowder was introduced into mines for blasting it was essential to remove all traces of firedamp first or risk disaster. The job of getting rid of the explosive gas went to the fireman, also known as 'the penitent' because of his costume. He covered himself in a hooded sackcloth garment, which he also soaked with water. He then crawled to the section of the mine where gas had been found. Once in place, he lay flat on his stomach, attached a lighted candle to the end of a long pole and then thrust it forward into the pocket of gas. The theory was that the flames from the resulting explosion would pass over his body and the wet sackcloth would protect him from burning. It was a theory that few were eager to put to the test. It was nevertheless not an adequate answer to the growing problem. Things became so bad that many miners preferred to work in the dark rather than risk explosions, but the situation was getting worse as mines went deeper.

The scale of the problem can best be illustrated by the appalling statistics gathered by an Inspector of Mines, Matthias Dunn, listing accidents in Durham and Northumberland between 1790 and 1840:[3]

	No. of Accidents	Deaths
Explosions	87	1,243
Suffocation by gases in the pit	4	18
Inundations from old workings	4	83
Falling of earth, rubbish, &c.	15	33
Chains or ropes breaking	19	45
Run over by trollies or waggons	13	12
Boilers bursting	5	34
	147	1,468

Explosions accounted for 60 per cent of all accidents and 85 per cent of the deaths. It was a situation that was steadily deteriorating throughout the latter years of the eighteenth century, but it took a calamity in 1812 to stir anyone into meaningful action. That year there was an explosion at the Felling Colliery on Tyneside.

A local clergyman, Rev. John Hodgson, wrote an account of the disaster in the local paper, the *Newcastle Courant*, much against the wishes of the mine owners who preferred to keep their disasters to themselves. He went on to preach a sermon on the disaster, which was also published, as was a fuller account as a booklet.[4] It is worth quoting at length for two reasons. Firstly, because it is rare to get such a detailed account of this type of accident, and secondly because the dramatic way in which Hodgson portrayed the tragedy stirred the public conscience and was directly responsible for bringing changes to the industry. After describing the nature of the colliery, and its two pits, he went on to describe the accident itself:

About half past eleven o'clock in the morning of the 25th May, 1812, the neighbouring villages were alarmed by a tremendous explosion in this colliery. The subterraneous fire broke forth with two heavy discharges from the John Pit, which were, almost instantaneously, followed by one from the William Pit. A slight trembling, as from an earthquake, was felt for about half a mile around the workings; and the noise of the explosion, though dull, was heard to three or four miles distant, and much resembled an unsteady fire of infantry. Immense quantities of dust and small coal accompanied these blasts, and rose high into the air, in the form of an inverted cone. The heaviest part of the ejected matter, such as corves, pieces of wood, and small coal, fell near the pits; but the dust borne away by a strong west wind, fell in a continued shower from the pit to the distance of a mile and a half. In the village of Heworth, it caused a darkness like that of early twilight, and covered the roads so thickly, that the footsteps of passengers were strongly imprinted in it. The heads of both the shaft-frames were blown off, their sides set on fire, and their pullies shattered in pieces; but the pullies of the John Pit gin, being on a crane not within the influence of the blast, were fortunately preserved. The coal dust, ejected from the William Pit into the drift or horizontal parts of the tube, was about three inches thick, and soon burnt to a light cinder. Pieces of burning coal, driven off the solid stratum of the mine, were also blown up this shaft.

As soon as the explosion was heard, the wives and children of the workmen ran to the working pits. Wildness and terror were pictured in every countenance. The crowd from all sides soon collected to the number of several hundreds, some crying out for a husband, others for a parent or a son, and all deeply affected with an admixture of horror, anxiety, and grief.

The machine being rendered useless by the eruption, the rope of the gin was sent down the pit with all expedition. In the absence of horses, a number of men, whom the wish to be instrumental in rescuing their neighbours from their perilous situation, seemed to supply with strength proportionate to the urgency of the occasion, put their shoulders to the starts or shafts of the gin, and wrought it with astonishing expedition. By twelve o'clock, 33 persons all that survived this dreadful calamity, were brought to day-light.

Altogether ninety-two died in the disaster, the youngest of whom was a boy of 10. The horror of the events produced an immediate response and a number of prominent local citizens got together and formed the Society for the Prevention of Accidents in Coal Mines, with Hodgson as one of the committee members. One of their first actions was to set in motion the search for a safe method of lighting mines, and they invited the leading chemist of the day, Sir Humphry Davy, to visit the north-east. He took samples of the gas and analysed them back in London with the help of his assistant Michael Faraday. For the first time firedamp was identified as being methane. Davy at once set to work on designing a suitable lamp. It was based on his discovery that the gas could not pass through a fine wire gauze surrounding the flame. He was not, however, alone in his work.

The engine-wright of the Killingworth Colliery on Tyneside had very little formal education, but from a very early age he had shown a great deal of mechanical ingenuity; his name was George Stephenson. He is mainly remembered by the title he was given later in life, the Father of the Railways, and his work on mine safety has been largely forgotten. He had no knowledge of theory but he knew mines. He decided that the first step would be to discover how the methane burned, and the only way to do that was to go down the pit and find a 'blower', a fissure that was leaking gas. To the considerable consternation of everyone else down the pit at the time, he held a candle to the blower and noticed that the gas tended to burn round the base of the flame and not round the top. He decided, on the basis of this experiment, that if he could increase the velocity of the draught of air to the flame, it wouldn't ignite the gas. He devised a lamp in which the air to the flame passed through a single tube, and he placed the device in a tall, glass chimney. He was convinced his device would work and he only knew one way to test it. In the company of the colliery's head viewer and friend Nicholas Wood, and the under-viewer John Moody, he set out to find the worst and most gaseous part of the mine. Moody later described the event:

> I accompany'd Mr. Stephenson and Mr. Wood down the A pit at Killingworth Colliery in purpose to try Mr. Stephenson's first safety lamp at a blower. But when we came near the blower it was making so much more gas than usual that I told Mr. Stephenson and Mr. Wood if the lamp should deceive him we should be severely burnt, but Mr. Stephenson would insist upon the tryal which was very much against my desire. So Mr. Wood and I went out of the way at a distance and left Mr. Stephenson to himself, but we soon heard that lamp had answer'd his expectation with safety.[5]

It was the act of a brave and self-confident man. He later made numerous modifications and in the end it was not very different from the more famous Davy lamp. The committee awarded Davy £2,000 for his work, but only paid Stephenson 100 guineas. There were many in the north-east, however, who regarded that as an insult to their local man and a subscription was begun that raised £1,000 for Stephenson. Davy was outraged: to him it was inconceivable that a revered scientist should share the honours with a mere miner. He could not accept that trial and error on the spot could produce the same results as careful observation and controlled experiments in a scientific laboratory. Davy wrote to Stephenson's benefactors asking them to withdraw their address of thanks to Stephenson, 'which every Man of Science in the Kingdom knows to be as false in substance as it is absurd in expression'. The committee ignored Davy. It must have been a source of great gratification to Stephenson to know that he was so highly valued by the people of his own community. The affair also left him with a lasting distrust of specialists, especially London specialists. The Stephenson lamp was christened the Geordie lamp and was to remain in use in the north-east, where it was considered at least as good as Davy's.

The two lamps marked a real step forward in mine safety, not only because they removed the risk of explosions caused by methane, but also because the colour of the flame provided an early warning that gas was present. Even in modern pits, where the electric helmet lamp has taken over the job of lighting for the miner at work, the safety lamps are still used as indicators. The lamps were not the only good to come out of the horrors of Felling. The committee also began pushing to have all accidental deaths in mines brought before the coroner, a move which was strongly opposed by the mine owners. It also marked a growing public awareness that there was often a heavy price to be paid for the coal in their grates. The message was soon appearing in broadsheet ballads:

> You men of wealth and luxury in county, town or shire,
> You seldom give a thought to us while sitting by your fire,
> Or think upon the dangers that threaten each poor soul,
> As fearlessly they go to work to hew and dig the coal.
>
> Down shafts ill-ventilated the miner he must go
> And crawl upon his hands and knees whene'er the roof is low;
> The hewer, putter, driver and the trapper in his hole,
> Are all exposed to danger whilst down among the coal.
>
> While in the dark and dreary mine, begrimed with dust and sweat,
> While thinking on sweethearts and wives and some dear household pet,
> 'The pit's on fire', that dreadful word sends terror to each soul;
> Overcome by gas, the miner meets his death among the coal.
>
> So just bestow a thought on us that labour down below,
> That work so hard by day and night to make your fireside glow,
> There is no harder working men, if you search from pole to pole,
> Than the honest-hearted miner who hews and digs the coal.[6]

If there were good messages to take from such an appalling event as the Felling disaster, there were also some more disturbing views being expressed. The mine owners resented every attempt to interfere in what they obviously still regarded as their personal fiefdoms. The miners had always fought a war against the enemies of gas and water, but there was another struggle that was to prove just as difficult: the battle between masters and men.

CHAPTER FIVE

MASTERS AND MEN

Industrial unrest has come to be seen as one of the defining features of the mining industry, but this was not always true of all types of mining. As mentioned earlier, rioting was all too common in Cornwall in the early years of the eighteenth century, but the anger was not directed at the mine owners. This lack of direct conflict was due almost entirely to the unique system of working the mines and paying the men. John Taylor, one of the most competent and adventurous of the mining engineers of the region, described the whole system in considerable detail, as he knew it at the start of the nineteenth century.[1]

At the top of the hierarchy were the landowners who leased the mineral rights to the adventurers, groups of men who put up the capital to start mining. The landowners received a royalty on the ore, but the day-to-day control of the finances lay with the adventurers. The adventurers appointed mine captains to oversee the actual running of the mines. They proved to be the key figures in developing the mines and were highly respected members of the community, as Taylor duly noted: 'It would be unjust not to notice here, how much of the perfection of the system of management in the mines has been owing to the zeal and intelligence of this respectable class of men, and how much in useful application constantly depends on their knowledge and activity.' The actual work of the mines was almost all carried out under contracts rather than by day rate. Taylor explained the advantages of the system from his perspective:

We now come to that part of the economy of the Cornish mines, which is most deserving of consideration from the effects it has produced, not only by procuring regularly a great deal of effective labour in proportion to the money paid for it, but also by turning that labour into such a direction as to make it the interest of the workmen to increase the discoveries of ore, and to work it and make it saleable in the most economical manner. Thus the owners of the mine have the advantage of the intellect and skill that the men collectively possess, and have only to guard against the chances of fraud which such a system may be supposed to be subject to, but which are in fact under intelligent and faithful agents of too trifling a nature to be accounted of any importance.

It sounds almost too good to be true, so how did this system operate? The work of the mine was put up to auction at regular intervals. Gangs of men got together and put in bids to do the work, the contract going to the lowest offer. Like many modern auctions, the auctioneer, in effect, had a reserve price and if that price was not reached then the mine captains would offer the work to anyone who would take it at that price. In theory, if no one came forward, the offer would be withdrawn and work would come to a standstill. In practice this never, or very rarely, happened. The contracts were offered for three types of work: tutwork, tribute and dressing. Tutwork was work that was done by measure, which included sinking shafts, driving levels or 'stoaping', opening up a section. Tribute was payment for winning the ore and dressing it to a certain level, so that its value could be ascertained; the men who did this were known as 'tributors'. Dressing was mostly extracting the ore from the waste left by the tributors, and was usually taken by a man who employed a team of women, known as 'bal maidens'.

The day at which the work was allocated was known as the setting, and several days before that the mine captains would have been making their own assessments of the likely costs and values of the different types of work on offer. The tutwork was put up first, and generally the men who had been working in that section before put up the first offer at a far higher price than they expected to get, to act as a deterrent to others. The winning group was known as a pair, which does not mean two men. The pair was split into corps of two or three men, who would alternate with each other so that work could go on, day and night. Once a price was agreed the men were provided with candles, gunpowder and any special equipment that would be needed, the cost of which was charged against their eventual payment. They were also given cash, known as subsist, for them to live on until the final reckoning at the end of the set period of weeks. It was usually the case that when all the bills had been paid, the men were still short of cash and so had no option other than to come back and work again at the next setting. Tribute work was more complex. Here the price received depended entirely on the value of the ore and the men took a certain proportion of the money – anything from a few pence in the pound to several shillings.

The system demanded a great deal of the miners. They needed to be able to assess the value of the work they were bidding for, as did the mine captains. Even in the more straightforward tutwork, they needed to be confident that they wouldn't hit any unexpected obstacles that would slow the work down or they could finish up out of pocket. Tribute was even more uncertain, as Taylor explained:

If a set of men working on a poor part of the lode where they may have agreed for seven or eight shillings in the pound, discover a bunch of ore rich enough to set at two or three shillings, they earn money very rapidly, and instances have often occurred where a set of miners have divided more than one hundred pounds a man for two months work. On the other hand, when the lode fails, and becomes poor, being obliged to go on with the contract, the men may at the end have their account in debt, not having even enough to pay for the articles they have consumed.

An all too typical set of accounts from the Devon Great Consols mine showed that the men had worked for two months and brought out ore valued at £182 2s 2d. They were receiving 7s 6d in the pound, which brought their earnings to £68 5s 9d. But out of that they had to pay £27 2d for goods they had received and repay £36 18s for subsist. That left just £4 7s 7d to share between the lot of them.

Under this system, the miners' earnings depended to a very large extent on their own skill and judgement, as much as on their ability to work hard. Normal shifts lasted eight hours, which did not include the time spent getting up and down the pit, but if they hit a particularly good lode the men would work as much as sixteen hours a day. They knew they only had the job for two months and they needed to make the most of their good fortune. It made for a very independent-minded workforce and a much more egalitarian society than was generally found in working-class communities. Everyone knew that they depended on each other's skills if a mine was to prosper.

> Dr Richard Couch, a mine surgeon in Cornwall in the 1850s, wrote: 'The active life of a miner, supposing it to commence at ten years of age, terminates at the very early age of twenty-eight, when, in most other occupations, he would be in the prime of manhood and vigour.'

All this did not mean, however, that everything was in some way idyllic in the copper- and tin-mining districts. A survey of death rates in Cornish mines compared with those in the coal districts of the north makes depressing reading. It showed for each decade of life, starting with 15–25 and continuing to 65–75, the death rate in Cornwall was appreciably higher.[2] Further investigation showed that the results were nothing to do with fatalities from accidents, but were almost all caused by respiratory illnesses. There was also a problem with 'diseases of the digestive organs', which were specifically stated not to have been caused by alcohol. The author, L.L. Price, who quoted those statistics, gave the official explanation:

Partly no doubt, it is due to the character of the food the miners habitually eat –
to its lack of nutritive qualities, and to the hindrance it presents to digestion; and
the fact was expressly noticed by the commission of the dyspeptic symptoms we
have mentioned above. Partly also it is due to the carelessness of the men, and
especially to the practice of racing up the ladders which is common among the
younger miners.

Price had his own, very different views. It was not racing up the ladders but
the fact that they were there at all as the only means of getting to and from
the work place, so that in some cases men spent as much as three hours a day
climbing up and down and traversing the long passageways. He also pointed
out that the working conditions were often atrocious, largely because so many
of the mines were wet, with water constantly underfoot and dripping from the
roof. One answer to part of the problem was obvious: introduce some form of
mechanical system in the shafts, such as the cages powered by steam engines
that were regularly used in the nineteenth-century collieries. But the mine
owners were reluctant to spend the money. One solution that was developed
in Cornish mines was the man engine.

In 1841 the Royal Cornwall Polytechnic Society offered a prize for the
best means of raising and lowering men in shafts. It was won by the engineer
Michael Loam, who built the first man engine and installed it at Trevesan mine.
A steam engine raised and lowered a set of stout timber rods, not unlike those
of pump rods, to which small platforms were attached at regular intervals.
There were also platforms at the same interval in the sides of the shaft. To get
to the surface, a miner stepped on to a platform on the rods at the bottom of
the down stroke. The next upstroke lifted him 12ft, where he stepped off on
to the fixed platform. He repeated this manoeuvre, stepping on and off the
engine, until he reached the surface. It was considered such a boon that it was
even celebrated in verse by the Gwennap poet, W. Francis:

> The engine by which he is raised from below
> Now supersedes climbing, health's deadliest foe –
> This miners know well and their gratitude show.
> Their core being o'er from labour they cease,
> And delighted avail them, O Loam of the ease
> Thy genius procured them and joyful ride
> On the rod, while others descend by their side.

The man engines may have been a huge improvement, but they were not
entirely accident free, though in many cases this was not due to any fault in
the engine. In May 1865 a miner tried to take two water kegs down with
him, which was strictly against regulations, and it seems that, as a result of
trying to manhandle the kegs, he missed his footing and fell to his death. That

August a young miner, who had used the ladders to descend the pit, decided to try the new machine for the return journey. But he had only gone up the first 12ft rise when he was paralysed with fear and was struck and crushed by the next stage. The main problems came from the accumulation of mud and clay on the stages, making them dangerously slippery, and on average there were two deaths a year throughout the district. Then, in 1919, there was one major disaster.

The engine at the Levant mine had first been installed in 1855, and gradually improved and extended over the years. By the time the accident occurred it was working down to a depth of 266 fathoms (1,596ft). The shifts were just changing over – one lot of men ascending and the others going down – when the rod broke near the surface, smashing through the safety catches that were supposed to hold it. The main damage was done in the top part of the shaft and it was here that thirty-one men were killed and eleven seriously injured. The remainder of the 150 men in the shaft at the time escaped with nothing worse than minor injuries. It was the worst disaster of its kind, and although investigations showed that the accident had been caused by a faulty iron plate, it brought the use of the man engine in Cornish mines to an unhappy close.

Another of the main problems facing the Cornish miner was poor ventilation in the mines, but the tributors had no option other than to follow the richest lodes, even if it meant getting further and further from the shaft. One miner in the St Just area described working in a level where the air was so bad that the only way to keep the candles alight was to have a boy wafting them with a fan. When the young lad fell asleep from exhaustion the candles went out and, as they were unable to relight them, they had to struggle back to the shaft in the dark. Conditions were at their worst after blasting. This account of the conditions was given by a 17-year-old working at Fowey Consols copper mine in 1842:

'The air', he said, was 'poor' where he then was and he had a pain in his head after working some time, which lasted for hours after he came to the surface. Almost every morning he had a cough and brought up some stuff black as ink. In the place where he was working they used to 'shoot' [blast] three or four times a day, after which they could not go into the end for half an hour, as it was full of smoke. He would eat his pasty in the level, where there was better air. Though he sweated a great deal and was very thirsty he could not generally get water underground.[3]

Bad air was not the only problem. The conditions met in different mines could vary enormously. One miner reported having to wade through icy water that reached his chest, while at the opposite extreme some men were forced to work in incredibly high temperatures. The mines of the St Day area were famous for this; the highest temperature recorded was at Wheal Clifford, which reached an incredible 125°F (58°C). There were other special conditions that occurred in different mines. Some Cornish mines, such as Botallack and Levant, had workings that spread out from the coast under the seabed. They were notoriously wet and the miners were often drenched with seawater, and, as a result, when their clothes dried out they became so brittle with salt that, as one miner put it, they were likely to break. The salt chafed their skin until they were red and raw. All this was bad enough, but when there were high tides combined with strong winds the water could flood into adits that opened out at beach level and rush into the workings. When that happened the men had no warning, and the best they could do was dash for the ladders; one man recalled many occasions when it seemed the water was rising faster than he could climb.

A very graphic description of Botallack mine was given by the novelist Wilkie Collins in his book *Rambles Beyond Railways*, whilst describing a visit to Cornwall in 1851. He had been brought to a point where he was far out under the sea when the miner who was acting as his guide told him to sit down and remain absolutely silent:

> After listening a few moments, a distant, unearthly noise becomes faintly audible – a long, low, mysterious moaning, which never changes, which is *felt* on the ear as well as *heard* by it – a sound that might proceed from some incalculable distance, from some far invisible height – a sound so unlike anything that is heard on the upper ground, in the free air of heaven; so sublimely mournful and still; so ghostly and impressive when listened to in the subterranean recesses of the earth, that we continue instinctively to hold our peace, as if enchanted by it, and think not of communicating to each other the awe and astonishment which it has inspired in use from the very first.
>
> At last, the miner speaks again, and tells us what we hear is the sound of the surf, lashing the rocks a hundred and twenty feet above us, and of the waves that are breaking on the beach above. The tide is now at the flow, and the sea is in no extraordinary state of agitation; so the sound is low and distant just at this period. But, when storms are at their height, when the ocean hurls mountain after mountain of water on the cliffs, then the noise is terrific; the roaring heard down here in the mine is so inexpressibly fierce and awful, that the boldest men at work are afraid to continue their labour. All ascend to the surface, to breathe the upper air, and stand on the firm earth.

There may not have been direct conflict between the men and the mine owners and adventurers, but there was, it seems, considerable complacency among the latter. In spite of the compelling evidence of many, many miners, the managers refused to accept that there was any problem with ventilation. In spite of inundations, it was cheaper to keep an adit open than install an expensive steam pump. One thing does seem very clear, the ill health and high death rate among the miners was not due to their carelessness, as the 'expert' so glibly explained. The unique system of organising work in the Cornish mines may not have been perfect, but it did have obvious advantages, not least in reducing industrial disputes to a minimum. In that respect it was very different from the collieries, where relations between miners and employers were frequently strained to breaking point.

During the first half of the eighteenth century, there was virtually no type of organisation among the miners. When disputes arose, they were either settled amicably or would suddenly erupt into a burst of rioting. Mines and mine buildings were the usual targets, and as drink was often involved many a landlord was liable to find his inn wrecked and his beer all gone. The results were usually the same: the Riot Act was read, troops brought in if necessary and ringleaders − or anyone unfortunate enough to be described as a ringleader − were arrested and sent to gaol. It sometimes seemed that trouble could start for no very obvious reason, as though pressure and resentment had built up, like steam in a boiler, until the safety valve had blown. At the heart of much of the trouble was the Yearly Bond that prevailed in Durham and Northumberland. Each year the miners would sign up to work for a particular colliery for the next year, in exchange for which they received a very small signing-on fee and an agreed rate of pay. It was a distinctly one-sided arrangement: the miners were to be available for work throughout the year, which was reckoned as a total of eleven months and fifteen days, but there was no obligation on the owners to actually provide that work during all, or indeed any, of that time. Needless to say, the owners were not required to make any payments to the men who were laid off. If, however, the men decided to take any other work to keep their families in food during those periods, they could be sent to gaol for breaking the contract.

Even if a miner managed to stay in work throughout the period, things could go badly wrong, and he had no redress. One of the most frequent problems arose from the system whereby the gangs of men were paid by the number of corves brought to the surface, but if any corve or tub was not completely filled it was valued as if it was completely empty − even though the coal that was in the tub would, of course, be sold and make a profit. The situation was particularly bad in some collieries where the 'keekers', the men given the job of assessing the corves, got a bonus depending on the number they rejected. One miner recalled working all day and sending up ten corves, of which seven were rejected. It was a bad system for the miners, and in 1765 it got even worse.

That year there was a shortage of hewers, so to prevent the men moving around looking for the best bargains, the owners introduced a new rule: 'No coal owner should hire another's men unless they provide a certificate of leave from their last master.'[4] As it was perfectly obvious that no owner was going to give permission to a good worker to leave for better pay, the miners would have been reduced to a form of servitude, forced to accept whatever conditions the owners chose to impose. On 25 August 1765, an estimated 4,000 miners came out on strike. The mine owners responded by issuing a statement that they forwarded to the Home Secretary, declaring that they would sign a 'certificate of leave' as soon as a man finished his full year's work. There was no indication, however, that this would be written into any formal agreement or be legally binding. The strike continued. The owners called on the Duke of Northumberland for help and he arranged for three troops of dragoons to be brought from York to patrol the mining districts, which did nothing to reduce tensions or calm tempers. Angry men attacked the pit at Pelton and set it on fire. In the end, an agreement was reached and the offending clause was dropped. It was the first victory for a concerted strike by the miners, but it did not end the Yearly Bond.

The next few decades were more peaceful and there was even a boom time for the miners, thanks to a shortage of labour. Matthias Dunn, who considered the bond a 'desirable arrangement' and opposition to it 'improvident', was horrified by the situation in 1804:

> A general scramble for hewers and putters took place at the ordinary bonding time. The fears of procuring the necessary supply of men were industriously magnified to such a degree that 12 to 14 guineas were given on the Tyne and 18 guineas on the Wear, and progressively exorbitant bounties were given to putters, drivers and irregular workmen. Drink was lavished in utmost profusion, and every sort of extravagance permitted. Nor did the evil end here; for a positive increase in all the rates of wages was established to the extent from 30 to 40 per cent.[5]

The hewers were the men who dug the coal, the putters filled the corves or tubs, and the drivers brought them to the foot of the shaft. The good times did not last, and soon the signing-on fees were reduced right back to less than £1. In 1809 the owners tried to consolidate their position by arbitrarily changing the binding date from October to January. The mines were always working flat out in October so the men were in a good bargaining position. In January work was slack and men were more willing to accept what they could get. It took a time for the significance of the change to sink in, but discontent grumbled on until the summer. In July, delegates from all the pits met and agreed to strike unless the change was reversed. The owners refused and the whole district came to a standstill in the middle of November. The reaction was even fiercer than it had been in 1765. The Combination Acts prevented any official union activity, but rumours were spread that men were taking 'oaths of brotherhood', Grand Guignol affairs that involved daggers through the heart and other fearsome forms of retribution for anyone who betrayed a brother worker. That was enough to justify military action: meetings were broken up by the troops and mass arrests made. The old gaol and the house

of correction in Durham became so overcrowded that some prisoners had
to be taken and locked away in the stables of the Bishop of Durham, under
military guard. Families were evicted from mine cottages and left homeless in
the depths of winter; they were not offered accommodation by any bishop.
With no official union and no funds for strike pay, the miners struggled on
for seven weeks, but eventually were starved into submission. The Yearly Bond
remained in various forms right through to 1872.

It was by no means the only grievance that profoundly affected the lives
of everyone in the mining communities. Indeed, few caused more bitter
resentment than the truck system, in which men were paid in goods instead of
cash, and the tommy shop, where the men were more or less forced by various
means to purchase their goods from the company shops. This chiefly operated
in the coal districts, but something very like it could also be found in the
south-west. The miners were required to buy their own candles from the mine
and it was generally reckoned that they cost a penny or twopence more than
they would have paid in the shops. It was not much but it brought an estimated
£400 a year extra profits to the big mines at the expense of their workers.
The situation in the collieries was even more serious. The tommy shops run
by the mines sold all the provisions the families might need. The men were
paid at the shops themselves and required to spend their money there; they
had no choice, simply because they were constantly in arrears in their wages
and in debt to the shops. As in Cornwall, prices were generally reckoned to
be higher than elsewhere and in some mines the men were not even paid
cash but were given tommy notes only exchangeable in the company shop.
There were various bodies that defended this system, including the Society
for Bettering the Condition and Improving the Comforts of the Poor, set up
by the philanthropist Sir Thomas Bernard and supported by others, including
William Wilberforce. They were no doubt well meaning in wishing to help
the poor, but they did not, it seems, trust working men to manage their own
affairs. In their report of 1796, they supported the tommy shop on the grounds
that 'the collier … is not able to squander the mass of his gains, to the injury
of himself and his family.' Apparently they believed that the system was being
run by kindly owners, concerned for the colliers' welfare and it was mere
coincidence that they made a handsome profit from the arrangement.

There were many who saw the obvious dangers in payment by truck and
tommy, and laws were passed against it being used in the woollen industry in
1726 and a stronger version appeared in 1749, but neither was applied to the
collieries. A more general act was passed in 1817, but there was still a major
problem in getting anyone in authority to administer it. Major breaches of the
act seemed to have taken place in the 1820s, but that did not necessarily mean
that the magistrates would enforce the law against the mine owners.

The troubles of the 1820s started when the iron masters and mine owners
of Staffordshire and Shropshire decided to impose a drop in wages of sixpence

a day. There followed the familiar pattern of sporadic rioting, followed by the calling in of soldiers and the death of two of the protesters. Eventually, agreement was reached and the reduction was settled at fourpence. The owners decided that they would make good the difference by raising the prices in the tommy shops, starting the process at Tipton, near Wolverhampton. A letter was sent to the Home Office, complaining about the events:

> The men have now availed themselves of a plausible and (I may Safely add) a real cause of complaint. By this practice the Coal and Iron masters compel their workmen to accept of two-thirds of their wages in goods, such as Sugar, Soap, Candles, Meat, Bacon, Flour, etc., instead of money at an unreasonable large profit. This appears the real cause of complaint more than the reduction of wages, and is really very hard upon them, and as the masters contrive to evade the Act of Parliament the men seem to have no relief but ceasing to work.[6]

The local magistrates met to consider the situation with the employers, without taking any evidence from the men affected, and came to the not surprising conclusion that really the miners were very well paid and that payment other than by cash was 'very rare'. They did, however, decide to enforce the law of the land by declaring that they would use the Vagrant Act against striking miners found roaming the streets. A familiar pattern of rioting and death was repeated. One magistrate, the Rev. A.B. Haden of Wolverhampton, stood out against the majority of his colleagues on the bench. He issued four summonses against coal masters who had been paying by tommy tickets, only cashable at one named store, and noted that 'It was curious to see what artifices they made use of, to prevent my fixing upon the most proper person to pay the just demands of the Complainants'. He was having none of it, and issued a stern warning that if the men were not paid in full the owners would be prosecuted in a higher court. It had the desired effect, and suddenly the right person was found to hand over the money. He added that 'The Court was crowded with Colliers, all of whom appeared much pleased with my decision'. I am sure they were: it was rare to find courts passing judgements that acted in favour of the men. Haden wrote to Robert Peel pressing him to strengthen the law to prevent the abuses in future, but without any immediate result. The laws were eventually strengthened and the tommy system stopped. It took many years of arguments to win justice for the miners and workers over this issue, but in the meantime other measures were rushed on to the statute books that were specifically designed to make it even harder for the men to fight the decisions taken by the masters.

CHAPTER SIX

THE FIRST UNIONS

The infamous Combination Acts effectively prevented miners and their families, like other industrial workers, from arguing for any change in their terms of employment. The acts had one great virtue: they were models of simplicity. Part of the act of 1800, for example, laid down exactly who would be liable for prosecution under the rules set out in the legislation:

> Every ... workman ... who shall at any time after the passing of this Act enter into any combination to obtain an advance of wages, or to lessen or alter the hours and duration of the time of the time of working, or to decrease the quantity of work, or for any purpose contrary to the Act, or who shall, by giving money, or by persuasion, solicitation or intimidation, or by any other means, wilfully and maliciously endeavour to prevent any unhired or unemployed journeyman or workman or other person wanting employment from hiring himself to any manufacturer or tradesman ...[1]

On and on it went, until the legislators were happy that they had covered every conceivable circumstance in which any workman could combine with others to improve their circumstances, or take any action to prevent others taking their jobs if anything went wrong. It was argued that, in the interests of fair and free trade, all such combinations were obstructive, not to say downright immoral. Anyone breaking the law as set out in the act would be brought before a magistrate – often a local employer – and could be sent to gaol for up to three months or to the house of correction. There were, however, no similar restrictions on employers: they could get together as often as they liked and agree to cut wages, increase hours or whatever else they fancied, and enforce the changes across a whole industry. The act was even extended to cover anyone who supported any combination in any way whatsoever. The first act of 1799 had been introduced by that great opponent of slavery in foreign parts, William Wilberforce, who saw no apparent contradiction between preaching freedom abroad and imposing draconian restrictions at home. These acts were, partially at least, all part of the reaction to events in France: the ordinary people were not to be trusted. Every one of them could become a revolutionary or Jacobin. There were some who went even further and claimed the authority of Scripture for what they were doing. Thomas Guest, the son of John Guest, founder of the powerful Dowlais Company of South Wales, no doubt voiced the opinion of many powerful owners when he pointed out that it was downright unchristian of men to oppose their employers: 'In providing for your own house you are not to infringe on the providential order of God, by invading the rights of others, by attempting to force upon those whom God has set over you, the adoption of such regulations and the payment of such wages as would be beneficial to yourselves.' As the popular hymn put it:

> The rich man in his castle,
> The poor man at his gate,
> He made them, high or lowly,
> And ordered their estate.

God had decided that Guest in his castle could reasonably ignore the poor, and they had no right to appear on his side of the gate. If the new laws were intended to ensure that Britain remained a peaceful place to live, untroubled by industrial strife, then they were an abject failure.

In the quarter century following the passing of the Combination Acts, life in many of the colliery districts was turbulent and often violent. Whatever the good intentions of the legislature may have been, the effects were soon all too obvious in many areas. South Wales was particularly hard hit as wages were steadily reduced, and many left their homes in the hopes of finding a living wage in other parts. The law may have prevented the men getting together to present their grievances, but that did not stop resentment spreading or mean that they would accept every dictate of the masters. By far the largest employer in the area was the Dowlais Company of Merthyr Tydfil, which controlled both collieries and ironworks, and they were among the most determined of all the employers to stamp out any attempt of opposition. It was not just wages that were cut: the company had always supplied cheap coal to the families of the men who worked the pits, but they now decided that was too generous a gesture and doubled the price of it. One man did speak out and the company decided that, as he was attempting to speak for others as well, he was acting illegally. The local magistrate agreed, so off to gaol he went. From there he had no choice but to write a wretched apology: 'I ham sorry that I abuesed your Honer in taking so much Upon me as to speak for Others ... Get me out of this whole of a place.'[2]

By 1816 the workers of South Wales felt that they had suffered all the wage cuts they could face, and so when Guest announced that pay was to be lowered even further they rebelled. Combination may have been forbidden but when they marched on Merthyr they were soon joined by the men from Dowlais. Guest rushed to sign up a band of men as special constables and armed them with pikes, but by the time they were organised the rebellion had grown too big to be contained, and his little private army was quickly disarmed. Guest hurried back to his home and when the crowd started to gather shots were fired from the house; several men were injured, one fatally. It was enough to turn the crowd away but not to stop them. More and more joined in what had become a procession, taking in all the local collieries and ironworks until there was an estimated 20,000 marching round the district. The magistrates called up the Glamorgan Regiment and it seemed that a familiar scene was about to be played out. But the magistrates drew back from confrontation and after lengthy discussions the decision to drop pay was reversed. It was a victory for most of the workforce, but those who had led the rebellion had to pay the price. They were promptly dismissed, and when they applied to the parish for poor relief were told they had to return to the parish in which they were born. They were to be put safely out of the way.

Failed by the legal system, the organisations went underground and took matters into their own hands. If the masters brought in blacklegs, then the miners were determined to drive them out again. A new, very secret organisation appeared in the valleys, known as the 'Scotch Cattle'. The origin of the name was obscure, but their objectives were clear. They would drive away anyone who tried to take work in the collieries during a strike. Organised into 'herds' led by 'the bull', they first appeared in Nantyglo in 1822. First of all bulls' heads were painted in a suitably gory red on the doors of houses where the blacklegs lived. If the warning wasn't heeded, then the herds of 200 or 300 men, all disguised and usually with blackened faces, would appear, rattling chains and ringing bells, and if that was not enough to frighten the occupants into submission, they could well receive a severe beating and see their property smashed. Soldiers were sent in to find the perpetrators and notices were posted all round the district offering a £15 reward for information. That was a huge inducement, but possible takers also saw another set of posters on the local walls. Painted in the same blood-red colours as the bulls' heads, they denounced all 'traitors and turncoats', and threatened to pull out the hearts of such men and stick them on the horns of the bull. No one offered any information to the authorities. The local press predictably condemned the Scotch Cattle when they emerged to play their part in a dispute over tommy shops:

We have to record further acts of outrage and violence perpetrated by those nocturnal yclept 'Scotch Cattle'. On Wednesday last week about 200 of these deluded men visited Blaencwm Colliery, distant from Pontypool about three miles and commenced in their usual manner, destroying furniture, etc., of the colliers and otherwise injuring the houses by hurling immense stones at them.

They have, unfortunately, succeeded in intimidating the peaceable and well-disposed from pursuing their labour. There are scarcely any collieries in the hills that have escaped their visitation, the most of which have been stopped working. It seems that the men are determined to be paid weekly for their labour in the coin of the realm, and will not be compelled, as before, to deal in the shops of their employers.

What the paper did not suggest was any alternative way in which the men could press what were, after all, more than reasonable demands.

Similar stories of thwarted ambitions were enacted in collieries all round the country. The owners used the Combination Acts as a big stick, with which they belaboured the leaders of any movement that attempted to negotiate terms. In Cheshire a group of miners met in secret to formulate a set of demands; however, faced by the threat of imprisonment they withdrew and were taken back to work, but only after making a very public and abject apology in which they had to 'humbly acknowledge the impropriety of our proceedings and return our thanks for the lenity we have experienced in the very serious prosecution that pended over us, being withdrawn'. Comparing that statement with the semi-literate plea made by the Welsh miner, it is not difficult to work out who wrote those words and presented them to the men to sign. Events such as this convinced many that they could not use reason to argue their case because the only outcome would be failure and humiliation. The alternative seemed all too obvious.

The men came out at the Radstock and Poulton collieries in Somerset in February 1817 and 'collected in a number of about three thousand and manifested some very serious symptoms of riot and destruction on the pits and the buildings annexed to them, which spread the greatest consternation through the whole neighbourhood'.[3] The Riot Act was read but the men took over the pits and vowed not to return to work until their demands were met. The next day, armed with clubs, they faced up to the cavalry with a battle cry of 'Bread or Blood'. There was only likely to be one winner. The usual arrests were made and within days the defeated miners were back at work. The account of the proceedings ended with a little homily to the workers to 'avoid those blasphemous and seditious publications which have cause their riotous conduct' and 'to look to their masters as their best friends'. The miners might have wondered why their 'best friends' had just cut their pay by 10 per cent. Striking miners at Dudley certainly didn't seem too friendly to their mine owner, Zephaniah Parker, when they threatened to hang him.

Sporadic rioting marked the first two decades of the nineteenth century. It was obvious that whatever the intentions of the Combination Acts might have been, they were not bringing any sort of peace, for disorders were erupting all over the country and not just among miners but throughout many different industries. Between 1824 and 1825 the acts were repealed. Many groups

The Combination Laws of 1799 and 1800 not only made it illegal for workmen to get together to demand higher pay or shorter hours, but also made it an offence to offer to help pay the defence costs of anyone who was charged under the acts.

There was no national miners' trade union in the nineteenth century, but different districts formed their own local unions following the repeal of the Combination Acts in 1824 and 1825.

of workers had managed to get round the legislation by forming Friendly Societies, ostensibly with the sole aim of raising funds to help members out if they became sick, injured or unemployed. The societies provided an opportunity for delegates to meet together and, once behind closed doors, to discuss whatever they chose. It is difficult to say exactly what went on as no one was going to keep minutes of illegal activities, but the organisations were there ready and waiting as soon as the law changed.

The Scots were first in the field, producing a pamphlet as soon as the first legislation went through.[4] It began with the resounding declaration that 'the remains of Gothic barbarism and feudal tyranny, known by the name of Combination Laws, have been swept from the Statute Book'. It then went on to give details of a meeting between delegates from twenty-seven pits in the region who had met and decided that 'it would be highly expedient to associate for the general good of the trade'. The union was quickly formed and was soon in action, attempting to change the system by which the amount of coal brought to the surface was measured – a measure that affected the price paid to the miners. The owners paid by what was known as 'a measure', which was not necessarily a fixed weight; previously it had been set at 26 hundredweight but it had crept up to 29 hundredweight. In other words, the men had to send up over 10 per cent more coal than before to earn the same amount of money. The demand for a return to the old system was sent to the owners. George Taylor, a local colliery manager, explained what happened next.[5] He refused to negotiate and told the men to go on strike, and to do so immediately. They did and he at once set about recruiting local Irish labourers to go down the mine. Soon production was up to half what it had been before the strike and some of the old workers began to turn up asking for their jobs back. They were taken on, but only after signing agreements not to have anything further to do with any union. It was not an encouraging beginning.

Things did not go much better at first in the north-east of England, where a co-operative was formed at Hetton, in Durham, by a miner called Mackintosh. Funds were collected but when questions were asked about what had happened to them Mackintosh suddenly disappeared and was next heard of in America. A much more promising start was made by another Hetton miner, Thomas Hepburn, in 1831. It was officially called the Pitmen's Union of the Tyne and

Wear, but more popularly as 'Hepburn's Union'. Hepburn became famous for his total rejection of the violence that had surrounded so many industrial actions in the past, and at a huge rally in Newcastle that year he was authorised to negotiate for, among other things, a reduction of the working hours for young boys from sixteen to twelve hours a day. When no agreement was reached, Hepburn called for a strike that eventually led to some concessions being made. It was a temporary victory, but as a result Hepburn became a full-time union official. The mine owners might have lost that battle, but now they were determined to win the war. At the next annual signing in 1832 they refused to employ anyone who was a member of the union. In spite of Hepburn's pleas for a peaceful settlement, violence broke out at Friars' Goose at East Gateshead.

The owners were using tactics that had worked well in the past. They began by evicting the striking miners from their homes, with bailiffs acting under police escort. Most of these were not regular police but special constables who had been sworn in for the occasion; one can safely assume they were not selected for their tact and discretion. At Friars' Goose the policeman in charge issued all the men with guns and, according to contemporary accounts, the evictions were carried out ruthlessly, with furniture being thrown out into the street and much taunting of the miners. Pleas for peaceful protest were always likely to be ignored in these circumstances. A number of men overpowered the sentry in charge of the house that had been taken over and used as the temporary police headquarters. They managed to take away some of the guns and ammunition intended for the bailiffs. At this point, the authorities decided to beat a retreat, but their way out led down a lane overlooked by a small hill. The miners on the hill started throwing stones at the police, who responded by opening fire on the crowd. The miners returned the fire and in the fracas five miners and two constables were injured before the police made their escape. The military were called for and eventually arrived accompanied by the Rector of Gateshead and the Mayor of Newcastle, but by that time the miners had dispersed. Examples had to be made, so forty people were arrested, including three women. The event went down in mining history as the Battle of Friars' Goose. It did not, however, bring the violence to an end.

On 11 June a South Shields magistrate was dragged from his horse and so severely beaten that he died from his injuries. William Jobling was charged with the murder, convicted and hanged, and his body hung on the gibbet at Jarrow for many weeks as a warning to others. Cuthbert Skipsey, a miner from North Shields, was trying to restore order during another disturbance when Constable George Waddle drew his pistol and shot him dead. Waddle too was tried for murder and convicted: he was sentenced to six months' hard labour.

At times there was a certain black humour in the situations that arose. A group of men broke into the house of a very unpopular overseer at Cowpen Colliery, made free with his larder and cellar, but left the rest of his belongings alone and his family unharmed. The next day they sent him a letter:

I see ye hev a greet lot of rooms, and big cellars, and plenty wine and beer in them, which I got me share on. Noo I naw some at wor colliery that has three or fower lads and lasses, and they live in won room not half as gude as your cellar. I don't pretend to naw very much, but I naw there shudnt be that much difference … ye maisters and owners may luk out, for yer not gan to get se much o yer awn way, wer gan to hev some o wars now.

But the masters did get their own way. Eventually, with money running out, the miners began to return to work and the strike petered out.

Thomas Hepburn had played an honourable part throughout all the disputes, continuing to argue against the use of violence by either side, but now he was banned from the coalfield and his work as a union organiser came to an end. He is still remembered in the area and every year a memorial service is held at his graveside at St Mary's church, Heworth.

Throughout the coalfields of Britain, the attempts to form legal unions were thwarted by employers, who used two powerful weapons: eviction of families from their homes and employing alternative, often unskilled labour as a stopgap measure until any strike was broken. In South Wales, the mine owners devised a new stratagem. They agreed that if any miner left a colliery he would be given discharge papers, stating why he had left, and that would include membership of a union. Anyone employing a new man would demand to see his discharge papers and union men were turned away unless they agreed to formally renounce their membership. If the men went on strike and the authorities considered there was a threat to the peace, they could call in the troops, read the Riot Act and arrest the 'ringleaders'. This was not a problem for the owners. When the men were in dispute with Thomas Guest of Dowlais, the sheriff gave the appropriate order: Thomas Guest was the sheriff. In this case things did not go quite to plan. The miners took matters into their own hands, attacked the soldiers, disarmed them and sent them scampering for safety. For a week it seemed like a famous victory as the miners held the field. It could not last. More troops arrived; the men, with no real organisation to fall back on, gave in. Attacking the military was a capital offence, and Dick Penderyn, the union leader, paid the price: he was hanged.

Throughout the 1830s there were sporadic strikes in all the British coalfields. In Scotland, these were mainly limited to Lanarkshire, where the truck system survived in spite of legislation. The law decreed that the men could not be paid in the company shop, but many owners got round this by simply putting up a thin partition in the shop to create a separate office. As in many other collieries, payments were generally fortnightly or even monthly, and many men were given subsistence to see them through while they waited for payday. They were then obliged to spend the money in the shop; failure to do so meant they were denied any further subsistence. Even those who were not in debt were under an obligation to use the shop. Those who refused found themselves transferred to the worst jobs at lower pay, or even dismissed. If the goods were not available from the company, the men were still not free to spend their money where they wanted: instead of cash they were given tokens. As a member of the Scottish Miners' Association explained:

> These tokens, however, have afterwards to be returned by the shopkeepers to the truck shop to be exchanged into cash, and the rate of exchange of 3s. in every 20s., in favour of the truck shop. This loss, amounting to 15 per cent., the shopkeeper has, of course, taken care has already fallen upon the miner.[6]

The system was reported as still being in place as late as the 1860s. The miners were not getting a fair deal for the money they did get, but from the 1830s there were a number of attempts to reduce wages, which led to more strikes by the fledgling unions. They were rarely successful.

By the end of the 1830s, attempts were made to provide a unified front of all industrial workers to press for a more democratic society. The People's Charter of 1838 called for universal male suffrage, a secret ballot, the removal of the need to have a certain amount of property in order to stand for Parliament, and payment for MPs. In brief: all men – but no women – should have the vote and it should be possible for anyone to become an MP. There was a wave of strikes in different parts of the country, including the mining districts, in support of the Chartists, but when the organisers declared August 1839 to be a 'sacred month' during which all the country's workers would come out on strike, the result was a fiasco. The movement for Parliamentary reform did not die with the death of Chartism, but in the long term it proved to have little connection with or understanding of the real grievances of workers. There were various reasons why Chartism failed to make a difference in the world, but there is more than a little truth in the view put forward by the early historians of the Trade Union movement: 'Made respectable by sincerity, devotion, and even heroism in its rank and file, it was disgraced by the fustian of its orators and the political and economic quackery of its pretentious and incompetent leaders whose jealousies and intrigues, by successively excluding all the nobler elements, finally brought it to nought.'[7]

By the 1840s the miners' organisations had become fragmented, pursuing purely local disputes. New unions were formed, only to be squashed again. The owners could use this situation to their own advantage. They could work together to keep the trade going, by shifting production from a striking colliery to another in a different district and by using a variety of devices to prevent men moving from one pit to another. Strikers were financially vulnerable, as the purely local unions never had any funds to maintain even a starvation wage to the strikers. It was either brave or foolhardy, depending on one's point of view, to go on strike. It was not just the men themselves who were penalised, but also their whole families, threatened with eviction from their homes. Not everyone felt able to face the prospect of ending up in the gutter surrounded by their few belongings, either smashed or tumbled into the mud. The owners knew that if things got really bad they could always call on the might of the state to intervene on their behalf. In the last resort, constables and cavalry would do their work for them. Perhaps the greatest bitterness was felt towards the men who came into districts to take on the strikers' jobs, a bitterness expressed in a ballad of the day:

> Off, ev'ry evening after dark
> The blackleg miners creep to work,
> With corduroys and coaly shirt
> The dirty blackleg miners.
>
> They take their picks and down they go
> To dig the coal that lies below
> And there isn't a woman in this town now
> Will look at a blackleg miner!
>
> They'll take your tools and clothes as well
> And throw them into the pit of hell
> It's down you go and fare you well
> You dirty blackleg miners.
>
> So join the union while you may,
> Don't wait until your dying day
> For that may not be far away
> You dirty blackleg miners![8]

It must have seemed that in this one-sided battle there would only ever be one winner, but just because the fights were so often lost, it did not mean that real grievances had been forgotten. For years, miners had lived in isolated communities, ignored by the rest of the world. All that was about to change, however, when the horrific conditions under which so many had been living and working were suddenly revealed to the whole of Britain.

CHAPTER SEVEN

WOMEN AND CHILDREN

I n Britain in the 1840s it was easier to find accounts of life among the
South Sea islanders or the native tribes of America than it was to read
about life in the mines of Britain. From the middle of the eighteenth
century there had been a deluge of travel books, covering, it seemed, all aspects
of life in this country, but only the most intrepid travellers ventured to explore
the underground world. This is not perhaps too surprising, since almost the
only time mines appeared in the news was to record either strikes and riots or
fatal accidents. The Rev. John Hodgson, who had done so much to promote
mine safety, did write an account of a visit to East Kenton Colliery, which was
connected to the Tyne by a 3-mile-long tunnel.[1] This was a very roomy affair,
6ft wide and 6ft high, laid with a single-track plateway, with occasional sidings
to allow the horse-drawn trams to pass each other. Hodgson described how
his party paused at one of these sidings while the boy in charge of the horse
stopped and called out, and then listened for approaching trams. They then
waited as another boy appeared out of the darkness, and the author writes
rhapsodically: 'The candle of the boy coming down appears like a star in the
distance, through the gloom, and has a very pleasing effect, as it gradually
approaches.' The whole trip sounds like a jolly jaunt, out of the ordinary
certainly, but free of any discomfort or danger. The author did, however, point
out that this system was unique and unlike the conditions down other mines,
but nothing was said about what the differences were.

One of the few authors who made a more demanding visit was Wilkie
Collins, who we met earlier enjoying the strange experience of walking out
under the sea in Cornwall. The real adventure, however, was in getting to that
point in the first place. The diminutive writer was fitted out with the standard
miners' protective clothes, which were all far too big – his trousers were pulled
right up to his armpits. This was all very well until they got to parts of the
mine where narrow planks led across deep shafts and reached one point where
a pit that disappeared down into darkness could only be crossed by widely
spaced beams. This was more than the baggy-trousered Collins could manage:

Our friend the miner saw my difficulty, and extricated me from it at once, with a promptitude and skill which deserve record. Descending half way by the beams, he clutched with one hand that hinder part of my too voluminous garments, which presented the broadest superficies of canvas to his grasp (I hope the delicate reader appreciates my indirectness of expression on the unmentionable subject of trousers!) Grappling me thus, he lifted me up as easily as if I had been a small parcel; then carried me horizontally along the loose boards, like a refractory little boy being borne off by the usher to the master's bitch.[2]

Very few travel writers welcomed the opportunity of being carried like hand luggage across an abyss. As a result, the world of the miner remained as terra incognita until the 1840s.

The mining communities may have been largely unknown, but the developing industrial world above ground could not be ignored. There was a growing interest in the textile mills and a deepening concern about the conditions of the young apprentices who provided a large proportion of the workforce. It was part of a more general movement to limit the hours worked in factories that led to the promotion of a Ten Hours Bill. That failed to get through Parliament, so another piece of legislation was brought forward, specifically relating to the children. It was met by horrified opposition, in part from those who regarded any interference by the legislature in private business as anathema, and also by the mill owners. Opponents claimed that reducing the hours that children worked would mean the ruin of the mill owners and an end to Britain's greatness. Among those who spoke for the bill was the great radical William Cobbett. He began his speech in Parliament during the debate by saying that he had always been told that Britain's glory had depended on all kinds of different bodies: the foreign trade of the merchant fleets, the stout yeomanry who tilled the land and even the banks that provided the money for enterprise:

But, Sir, we have this night discovered, that the shipping, the land, and the Bank and its credit, are all nothing worth compared with the labour of three hundred thousand little girls in Lancashire! Aye, when compared with only an eighth part of the labour of those three hundred thousand little girls, from whose labour, if we only deduct two hours a day, away goes the wealth, away goes the capital, away go the resources, the power, and the glory of England![3]

The bill became law as the Factory Act of 1833, which made it illegal to employ any children under the age of 9, limited the working hours of 9- to 13-year-olds to eight hours a day, and 14- to 18-year-olds to twelve hours. It also provided for inspectors to be appointed to ensure the law was not broken, though in practice they had only limited access to workplaces. The act did not apply to any industries apart from textiles, but it did set a precedent, and one of its most enthusiastic supporters and advocates, Lord Ashley, later the Earl of Shaftesbury, continued pressing for better conditions for children in work. Ironically, he was helped by the mill owners, who were complaining that it was unfair that they had been singled out for legislation when other industries could do what they pleased. The result was the establishment of a committee to investigate the conditions of children working in mines. The findings were published in 1842.[4]

The reports are extraordinary documents. The country was divided up into twenty-four different mining districts from Cornwall up to Scotland, and inspectors made detailed surveys of everything from working practices to the morality of the children. They went down mines to see conditions for themselves and interviewed hundreds of witnesses of all kinds, from mine owners and local magistrates to the young children. The most interesting sections come at the end, which detail the evidence given by the witnesses. Some idea of the thoroughness with which the inspectors went about their task can be gauged by the sheer numbers of people interviewed: the report on Northumberland, for example, provides evidence taken from 668 people. No one any longer had any excuse for not knowing what life in the mines was really like. The results even came as something of a shock to some owners. As one report noted: 'the coal owners seldom or never descend into the pits; few of them have any personal knowledge or take any superintendence whatever of their work-people.'

One advantage of the system of publishing separate reports was that it highlighted regional variations, and these turned out to be very significant. Many commentators have, quite rightly, railed against the facts that emerged on the employment of women and girls below ground, but it turned out that this was far from a universal practice. In fact, the number of collieries doing so were in a minority: in Glamorganshire, for example, the collieries employed 358 boys, but only 43 girls, and in some regions girls were never employed at all. Although the work was much the same in one colliery as in another, the methods of hiring and paying the workforce could vary enormously and different areas had quite different names for the different occupations and equipment. The trucks used underground to move the coal from the face to the foot of the shaft were variously known as trams, drams and corves. The children who pulled and pushed them were often referred to as 'hurriers', but in South Wales they were 'scooters'. In Northumberland there were fine distinctions, which affected the pay. Boys who worked alone were 'trams' and got a shilling a day. Sometimes they were helped by younger boys called 'foals' and the older boy was now the 'headman'. They split the money, with 4d going to the foal. If the two boys were the same age, they were 'marrows' and split the money fifty-fifty. Given such variations, it is easier to understand the difficulties organisers had in creating wider unions between miners. But, in spite of the differences, the overriding effect of reading this great mass of material is to get a feel for the human misery that was revealed, not as dry statistics, but in real stories told by real people.

The first problem all the commissioners had to overcome was gaining the trust of the mining community. John Roby Leifchild, who was responsible for investigating conditions in Northumberland, wrote: 'For a stranger to read the mind of a pitman, a circuitous approach and no small tact are requisite.' The miners were deeply suspicious, mainly because any survey that had been conducted in the past had never turned out to their advantage. 'It would seem to be assumed as a truth, amply established by experience, that his master can have no desire to benefit him and any expression of such a wish appears to him the mere forerunner and disguise of an unprofitable proposition.' Inevitably, they had similar doubts about the commissioners: 'the rulers of the land would not have contrived a mission to their colonies and an examination into the minutiae of their labour and their gains, without a view to the imposition of a tax.' But the information was gathered and it would seem fairly and faithfully reported. Leifchild also made the important point that for readers of the reports to understand the language used, it had to be seen in context. When a child said his or her work was easy, it did not necessarily mean that it was easy in any objective sense, just that some parts of the work down the pit were even harder. The younger witnesses 'were unable to transfuse the bitterness of their thoughts into their homely terms and phraseology': a boy when asked whether he had ever had any accidents replied that he had once hurt his arm. This could be interpreted as a nasty bruise, but it had been fractured. Even bearing this in mind, the stories that the children have to tell are harrowing.

The age at which children started work varied from district to district. In Lancashire, some were 'so young they go in their bed-gowns'. In South Wales, trapper boys were working down the pit at 5 years old. The long, tedious hours working in the dark were difficult to endure and the very youngest found it difficult to keep awake. One boy, who was 7 when he gave evidence, said that he had been down the pit for three years: 'When I first went down I couldn't keep my eyes open. I don't fall asleep now, I smoke my pipe.' The task was tedious, simply opening and closing the doors, using a length of string or rope,

The Royal Commission Reports on Children in Mines was published in 1842. Britain was divided up into twenty-four different regions, and inspectors in each region visited the mines and interviewed everyone from owners to young children. The country was shocked by the revelations of what the children endured.

The Mines Act of 1842 made it illegal to employ females of any age and boys under the age of 10 in underground workings.

and there was nothing there for their comfort. The children sat in a scooped-out hole in the ground, usually in the dark. One of the commissioners, visiting a Yorkshire colliery, found children as young as 5 at the job and told this sad tale:

> On one occasion as I was passing a little trapper, he begged me for a little grease for my candle. I found the poor child had scraped out a hole in a great stone and having obtained a wick, had manufactured a rude sort of lamp and that he kept it going as well as he could by begging contributions of melted tallow from the candles of any Samaritan passing by.

The reason that the children had to work in the dark was simple: it was cheaper. The trapper would have been charged 1½d out of his 10d pay for the use of candles.

It is difficult to imagine how such young children could keep awake, sitting in more or less permanent darkness for hours on end. Joshua Stephenson was one of those who had started work at the age of 5. He left his home at three every morning and didn't get back again until five in the afternoon. When he did return, he only had enough energy to eat his evening meal and then fall into bed, ready to set off again the next morning. Yet these youngsters were a vital part of the safety of the pit. It was their job to ensure that all the ventilation channels were kept open to prevent a dangerous build-up of gas. Failure could lead to catastrophe. That is exactly what happened at Wellington Pit in Northumberland on 19 April 1841. The commissioner was there at the time and was able to take the evidence of John Johnson, who led the rescue attempt. He described the scene as he made his way down to one of the seams at a depth of 140 fathoms. Accidents had been reported in the press in the past, but his was something very different; a first-hand account of the horrors that waited to greet the would-be rescuers:

> It was some minutes before our eyes became accustomed to the horrible gloom, although groans were heard from some of the poor sufferers; at the bottom of the shaft we found two lads, one an onsetter, the other a rolley-driver, both of them alive; they were sent to the bank but survived a very short time. Lying near to them two other lads quite dead; all these poor fellows were dreadfully burnt. My first object was to examine the way to the west but we had not proceeded more than six or seven yards we came to an immense fall of the roof, occasioned by the timber that supported it having been blown away. Near these falls were lying a horse and pony, severely burnt and mangled, they were quite dead. We continued west for forty yards; we were met with the after-damp, but hearing a groan up the north headway, we rushed up for fifty-six yards and succeeded in bringing out a body, but being much burnt he could not be known. He died after being taken to bank, where he was recognised as James Pearson, a putter.

Having persevered for this distance without air in the after-damp, we had great difficulty turning, feeling quite sick and giddy. In our way up the headway we found that the stoppings were blown out and that consequently the current of air had ceased to ventilate the entire west and north workings and that all the people who were in that part the pit, who might have escaped the effects of the explosion, would lose their lives by the after-damp.

He was forced to close off part of the mine to prevent further explosions that would have killed the rescue party. Once that was done, the gas could be vented off and the rest of the bodies removed. They found the body of a trapper boy called Cooper some distance from his door, and decided that he must have left his post to chat to another boy, whose body was found near his. Without the proper ventilation in place, a miner must have come down with a lighted candle with deadly results. The disaster, which claimed thirty-two lives, was attributed to the 'gross negligence' of the boy.

As Cooper was vilified for his fatal neglect of his duties, it seemed no one was ready to question those who had entrusted the safety and lives of the mine to an 8-year-old boy. You can never make direct comparisons of lives today with those of the nineteenth century, but anyone who has ever had anything to do with primary school children will know how easily they get bored and how difficult it is to get them to sit still for any length of time. It is not hard to imagine that little boy, alone in the dark, sneaking off for a few minutes for a bit of human company. And Cooper, at 8, was already getting quite old to be a trapper; many children of that age were being moved on to more demanding tasks.

The author of the report added his own opinion on the explosion and its aftermath: 'But the excitement gradually subsides and despite of reports, letters and observations, no alteration is deemed by viewers possible or advisable and boys of 7, 8 and 9 years of age still continue to tend the doors.' If anything, this assessment was generous. One little trapper in a nearby colliery 'calls himself 6 years old', though he wasn't sure of his age. He got up at three every morning and an hour later he was down the pit. Lives of men depended on the vigilance of little boys like him, dragged from their beds before dawn to spend ten hours or more underground. Also, in a small minority of mines, the job was given to little girls. Why were the children employed at such a young age? The answer given by many of the employers was 'the greed of the parents' – they wanted the extra money: 'At five years of age, however, arise the call for labour. To add an extra half-crown per week to the wage of the father, he would be sent into the mine. From this period the mother seems to consider herself relieved of the responsibility for the treatment of the child.' They followed the example of their fathers: 'swearing, drinking and obscene language are all too frequent among the young.' By the time they were 8, many children were running up bills for tobacco at the shop.

Once the children were considered big and strong enough, they became part of the team whose job it was to move the coal from the face to the bottom of the shaft. In most mines the coal was moved in wheeled trucks, but in a few more primitive pits it was either carried on people's backs or dragged along in baskets. Already by this time some collieries were using horses and ponies to haul the trucks, with the children acting as drivers. This might seem like quite light work and there is a good description of what was involved in the preamble to the report on mines in Glamorganshire:

> The duty of the haulier is to drive the horse and tram, or carriage, from the wall-face, where the colliers are picking the coal, to the mouth of the level. He has to look after his horse, feed him in the day and take him home at night. His occupation requires great agility in the narrow and low-roofed levels. Sometimes he has to stop the tram suddenly. In an instant he is between the rail and the side of the level and in almost total darkness, slips a sprig between the spokes of his tram wheel and is back in his place with amazing dexterity, though it must be confessed, with all his activity, he is frequently crushed.

The dangers were always present and there are numerous accounts of broken limbs, caused by carts running over them. On the other hand, they had a better life than many others down the pit: 'As a class these youth have an appearance of greater health than the rest of the collier population … with fair animal spirits and on horseback, going or returning from work, galloping and scrambling over the fields or road, bear the aspect of the most healthy and thoughtless of the colliery boys.' It seems almost pleasant, ending a day galloping across the fields, but because the work was easier than other jobs in the pit, the boys could be kept at it even longer. One boy was kept down for three days without once coming to the surface and without sleep, but said he didn't mind because he needed the money.

Horses were almost entirely used, as here, in drift mines, where they could be led straight into the mine. Elsewhere, the job of moving the coal fell to women and children. Eleven-year-old Henrietta Frankland worked near the mine where the horses were used as a 'drammer', hauling the carts or drams. Each dram held from 4–5cwt of coal and she had to drag it along roads that were very wet and only 30–33in from floor to roof. She made up to fifty trips a day, but at the time of the interview she had been off work for two months: 'a horse fell upon me and the cart passed over me and crushed my inside.' Another boy reported that the trams 'are so many and the road is so crowded, I have broken my arm three times and my leg once'. Others worked in even worse conditions, where the roof was no more than 2ft high. Here they were forced to wear a harness, which was attached to the tram by chains. They then crawled on all fours, dragging their loads behind them. At Shilbottle Colliery in Northumberland, the boys worked in 30in-high seams, and pushed the trams with their heads. They took off their stockings, wrapped them in their caps and used that as a cushion for their heads. Even so, one girl showed her interviewer a bald patch, which she had got from working this way for many years. Many of these children had no meal breaks. One boy, who worked with another lad, described how he managed to grab something to eat while sitting in an empty truck, pushed back by his friend. Another described how he frequently had to leave his food until he got a chance to eat it, only to find that the rats had got there first.

The harshest treatment seems to have been handed out to pauper children, sent to the mines by the parish poor house. William Forrest was only 8 when he was first harnessed to a cart. He worked in a place that was always knee deep in water and sometimes up to his belly, and he crawled along balancing on his hands and feet: 'Sometimes when he was tired he had to stop and then he was always licked with a strop.' Other times he was beaten with a pick handle. The other children had family around them, which offered some protection from abuse; these unfortunates had no one to care for them. All these children had hard lives, but none suffered more than the women and children of Scotland.

Robert Hugh Franks, who reported on the mines of east Scotland, wrote a graphic description of the conditions that he found:

> The roads are most commonly wet, but in some places so much so as to come up to the ankles and where the roofs are soft, the dripping and slushy state of the entire chamber is such that none can be said to work in it in a dry condition, and the coarse apparel the labour requires absorbs so much of the drainage water as to keep workmen as thoroughly saturated as if they were working continuously in water.
>
> The workings in the narrow seams are sometimes 100 to 200 yards from the main roads, so the females have to crawl backwards and forwards with their small carts in seams not exceeding 22 to 28 inches high.
>
> The danger and difficulty of dragging in roads dipping 1 foot in 3 or 1 foot in six may be more easily conceived than explained and the state which females are in after pulling like horses through these holes, their perspiration, their exhaustion, and very frequently even their tears, it is painful in the extreme to witness.

Eleven-year-old Janet Cummings told her own story. She went down with the women at five in the morning and came up again at five in the afternoon, except for Fridays when she worked all night and didn't emerge until noon the following day. She carried large pieces of coal from the face to the shaft, but smaller pieces were loaded into a basket, which she carried on her back. She said she carried a hundredweight; she had no idea how much that was 'but it is some weight to carry'. Sometimes she would have to take her load for a quarter of a mile: 'The roof is very low and I have to bend my back and legs and the water comes frequently up to the calves of my legs. I have no liking for the work, father makes me like it. I have never got hurt, but often am obliged to scramble out of the pit when bad air was in.' Agnes Moffat was one of the girls whose job was to take the coal up ladders to the surface. The basket was held in place by a headband and she described the frequent accidents when the bands broke and showered the coal on to the girl below. She summed up her own feelings: 'The lasses hate the work altogether and canna run away from it.' The effect of all this desperately hard work was explained by Jane Peacock Watson, who was 40 and had been working in the mine for thirty-three years. She had been married for twenty-three years, during which time she had given birth to eleven children. Two were stillborn, three died of typhus and the rest were still alive. She always worked right up to the time of the birth and was usually back again after ten or twelve days, sometimes less if she was needed. She said that stillbirths and miscarriages were common 'from the oppressive work' and it was not unknown for women to give birth down the mine, and carry the baby out wrapped in an apron. They were all old women at 40.

Children did do other jobs in the mines, such as helping to work manual pumps to bring water up to the level where it could be brought to the surface by the steam pump or working windlasses to haul trucks up steep slopes. One girl of 19 had the demanding task of filling the trams, working with one miner. She also took on the job of drilling holes ready for blasting and even sometimes helped with hewing the coal. But these examples were few. Many children did, however, complain of being made ill by the conditions underground. Joseph Beaney's case was, unfortunately, all too typical:

> After he was down 6 months he fell very sick and had headaches, bringing up his food off his stomach sometimes thrice a week. Has been in bed ever since: sometimes worse, sometimes better. Sometimes now cannot eat his breakfast of mornings. Has been sick and spewed up this last fortnight. Thinks the air down the pit is cause of this.

Some doctors who gave evidence corroborated the effects of the harsh life on the health of the miners. Dr S. Scott Alison gave his views of the effect of the work on the Scottish miners: 'Ere these children have been many years in the collieries death has thinned their ranks.' Accidents were common and he described fingers and toes being lost or so badly mangled as to require amputation. Coal dust got in the eyes, causing blindness, and breathing difficulties were common. Few were in good health by the time they reached 20, and in the next few years the effects of respiratory diseases became more severe. There was a steady deterioration and by 40 'the symptoms of decay now succeed fast, and death is busy with the selection of his victims'.

The reports did not just look at working conditions, but also described the lifestyles and living conditions of the mining communities. Housing varied enormously from district to district. In Durham, for example, Coxhoe, a typical mining village, consisted of terraces of ten to twelve cottages, soundly built of stone rendered with lime plaster and with slate roofs. Downstairs was a front room, roughly 14 by 15ft, a back room 14 by 10ft and a small pantry. Upstairs was a single bedroom right under the roof, so that the vertical walls were just 2ft high. Because the women in the Durham coalfield never went down the mines, the houses were always neat and tidy, and though they had no gardens, many had allotments, where they grew vegetables and often flowers. The miners were well known for carrying off the prizes at local flower shows. Home life in Lanarkshire could hardly have been more different:

> The hut itself is a wretched hovel, perhaps 10 to 12 feet square, in which a family of from six to ten individuals are huddled together, two bedsteads and sometimes only one, nearly destitute of covering, generally a few stools, sometimes a hanging of a chair, and some damaged crockery, fowls, occasionally a pig or a jackass, dogs, and whatever animals it may chance that they possess, share the room with the family and the only objects of comfort which present themselves are the pot, and the fire over which it invariably hangs ... There is generally an absence of all drainage and the filth &c., of each cottage is accumulated *before* the door, not even, in many cases, placed on one side. Indeed, there is rarely any other deposit for filth except the entrance to the dwelling and even this filth itself is not neglected as a source of profit. One of the witnesses informs us that his father said that dung and filth paid for the whiskey, and I believe the purchase of whiskey is usual destination of the profit of the abominable and unhealthy nuisance.

One witness said that the lack of furniture made it easier for the family to 'flit', to leave the area to find different work. Inevitably, the vile conditions were put down by the owners as due to intemperance, but witnesses made it clear that the main reason was the state of exhaustion of the women when they got home after their gruelling work.

One group that escaped censure was the lead miners of Durham. They were nearly all literate and libraries thrived. There was a very strong temperance movement, and many men went to the local pub for the company and spent the evening smoking their pipes, drinking non-alcoholic beverages and enjoying the conversation. This saddened the landlord: 'What a pity it was that a poor man should be deluded to spend their money on such stuff when he might get good wholesome, invigorating beer.' It would seem to be the only time in all the reports where anyone criticised the miners for not drinking. A rather more typical response was made by a miner when asked how many drunks there were in his house: 'Will you count them as I call their names. There's my son, John, Jim and William, Dick, Thomas, Ned and Joe, that makes seven and there's myself,' and, pointing to his wife, 'There's the old woman, you may put her down for she gets drunk as well as the rest of us.'

Although the men of the lead-mining district had good home conditions, they were not always able to enjoy them. Many of the mines were a long way from the villages, high up in the dales and rather than trudge many miles to and from the mine each day, the men stayed four nights a week in lodging houses for which they paid 1s 6d a week. One of these was a barn-like two-storey building. The ground floor was just one big room with long tables and benches, lockers for the men to keep their provisions and a fire at one end where they could heat their food. A ladder led up to the floor above, which contained fourteen bunk beds that could sometimes sleep as many as forty men and boys. It was said that the best feature of this stark accommodation was the stream of crystal-clear water that ran past the door.

Morality was one of the aspects of miners' lives that had a section all to itself in every report. There was a great emphasis on education and in particular religious education, or rather the lack of it. The commissioners were horrified by the fact that many of the children had only the vaguest idea of Christianity. One child described Jesus as the son of Adam, and knew no more than that, and another said she thought she had been made by a man in the sky but didn't know who he was. There were frequent mentions of the fact that even when Sunday Schools were provided very few attended. Perhaps the most honest answer to why this was so was provided by Ann Egley: 'I went a little to Sunday School but I soon gave it over. I think it too bad to be confined both Sundays and week days. I walk about and get the fresh air on Sundays.' Who can blame her?

When the reports were published there was a real sense of horror at what was revealed but, far more than the accounts that were given of harsh labour conditions, long hours and wretched homes, it was the stories of immodesty that seemed to create the greatest scandal and which received the widest publicity. Readers did not even have to plough through page after page of evidence to find just how shocking the situation was. All they had to do was look at the illustrations. One that caused particular outrage showed two teenage children, William Dyson and Ann Ambler, being wound up from a shaft. They were sitting on a narrow iron bar, attached to a chain that was wrapped round a windlass being worked by an old woman, much as if they were buckets being drawn from a well. It was not the inherent danger of the situation that caused concern, but the fact that the boy and girl had their arms wrapped round each other and they were both naked above the waist. An almost equally disturbing illustration showed a teenage girl half naked, on her hands and knees and harnessed to a tram. Alarmed by such sights, readers could find anecdotal evidence to go with the illustrations, much of it from Yorkshire pits. One of the inspectors found a group of men, boys and girls:

… some of whom were of the age of puberty, the girls as well as the boys stark naked to the waist, their hair bound up in a tight cap and trousers supported by their hips ... Their sex was recognisable only by their breasts and some little difficulty arose in pointing out to me which were girls and which were boys, and which caused a great deal of laughing and joking.

The miners seemed less concerned than the inspector, but 17-year-old Patricia Kershaw was obviously quite disturbed by her surroundings: 'The getters that I work for are naked except for their caps. They pull off their clothes. I see them at work when I go up ... the boys take liberties with me sometimes, they pull me about.' The men were not being deliberately provocative – they worked naked because of the extreme heat and humidity in the mine – but it is easy to see that this could be disturbing for a teenage girl.

In among the stories of deprivation and misery, there were regular comments from the owners and their representatives. The proprietor of a mine in Pembroke remarked: 'I think a limitation of age would be a barrier to their being brought up to working habits.' In Yorkshire, a number of owners got together to form a committee of their own and sent in their findings. They labelled the suggestion that there should be any form of inspection as 'espionage', described any reduction in hours as unworkable and declared that if young children were prevented from working they would all be ruined. They ended with this brazen statement: 'Your Committee deny that there is anything in the nature of the employment in coal and iron mines that affects the health.' Some owners, to their credit, took a very different view. It would have been difficult for individual owners to make changes to their own working practices and remain competitive, but the Duke of Buccleugh's manager was firmly in favour of legislation:

If a measure were passed enacting that no female were to be employed in our pits at all, no boys allowed to go down under 12 years of age, and only then if they could both read and write and in all cases the work limited each day to 10 hours, if such a measure were to pass I do not know a greater boon that could be conferred, not only upon the mining population, but upon the proprietors of Scotland.

Public opinion was moving towards the duke's viewpoint, and Lord Ashley introduced a bill to Parliament. He had considerable difficulty getting anyone to sponsor the bill in the House of Lords, where many of the big landowners got a great deal of their income from their mineral rights. In the event the bill was passed, but only after a great many amendments. Ashley remarked that the lords had 'left the bill far worse than they found it'. The Mines Act was passed in 1842, but its provisions fell a long way short of the ideals expressed by the Duke of Buccleugh. It did decree that no females should ever be allowed to work underground and that boys could not go down before their 10th birthday. There was, however, nothing in the act about working hours and apprentices were still allowed to be sent from the poor houses. Inspectors could be appointed, but they were given no authority to investigate the actual working practices, only to report on the 'condition of the workers'. Nevertheless, it was a huge step forward, though not everyone thought so. When women lost their jobs below ground it had an immediate effect on household budgets: poor people had just got poorer. Even those who found surface jobs were not always satisfied. The 'pit-brow lasses' of Lancashire even preferred the old ways. They 'liked it reet well – would like well to work below again – liked it better than working up here'.[5]

The Mines Act of 1842 righted some wrongs, but there remained many safety issues that were ignored, and it did nothing to help the mining community as a whole achieve their ambitions of getting fair pay for a reasonable amount of work.

CHAPTER EIGHT

SEARCHING FOR SAFETY

Popular histories of mining in nineteenth-century Britain have a simple story to tell. There used to be fires and explosions in mines, then along came Davy in 1815 with his safety lamp and the problem was solved. Sadly, this was extremely far from the truth. The lamp should have made a huge difference, and could have done had it been used correctly, though it was never to provide more than a partial answer to the whole question of mine safety. Between the years 1835 and 1850, well after the introduction of the lamp, there were 643 explosions in the north-east of England alone. That works out at an explosion happening somewhere almost every single working week. Amazingly, in collieries up and down the land, miners still worked by the light of naked candle flames. There were a number of reasons for this. The light from the safety lamp was considered too dim. Some even considered that it made the miner's life more dangerous. One witness, a man with forty years' experience in collieries, giving evidence to the 1842 commissioners, said that it 'has in practice been a curse, because it enabled men to work in badly ventilated mines'. He gave an example of men who 'left a colliery in the neighbourhood a short time ago on the ground that they could not work there with safety because of the air gate that had been stopped to save expense'. The owners' response was not to improve safety, but to take the men to court for breach of contract. There was, of course, another reason why the lamp was not universally used: cost. Instead of being used by everyone in the mine, the lamp was taken down to test for gas. If there was methane present the lamp burned brighter, with a bluish tinge to the flame:

> When the fireman (or his representative under another name) went his morning rounds, previous to the workmen entering the mine, his duty now usually only extended to examining every working-place to ascertain whether any fire-damp was present, a task in the performance of which the Davy lamp rendered invaluable aid; one or two lamps being commonly kept for this special purpose at mines where fire-damp existed at all, though otherwise worked with naked

lights. His mode of procedure differed in different districts. Sometimes he was required to place a lighted candle in each working-place found free from fire-damp, without which signal the workmen were forbidden to enter. In other cases, on gas being found present, he placed some obstruction in the way, such as a prop, or cross-sticks, or a shovel, or pulled up the rails, to indicate that the place was dangerous.[1]

It sounds a sensible system, but it was based on a number of assumptions: no major gas leak would occur after the morning inspection, gas leaks in a part of the mine not being worked would not seep through to other areas, and nothing would go wrong with the ventilation system. None of these assumptions was justifiable. As the account of the explosion at Wellington Pit in the last chapter made clear, the explosion may have been 'caused' by a failure in the ventilation system, due to the trapper boy not ensuring that the flow of air through the workings was maintained, but the gas was only set on fire because the miner was using a candle with a naked flame, not a covered lamp. Gas could appear anywhere at any time. At Seaton Delaval Colliery in Northumberland, in November 1843, a miner was working as usual when he literally hit a pocket of gas, which took fire from his candle and soon spread to the surrounding area. That part of the mine had to be closed off for six weeks, at the end of which the area was opened up again and was found to be still full of methane. It was actually measured at 48,000 cubic feet of gas, and all that from just one blow of a pick.

Detailed notes on mine explosion in the first half of the nineteenth century, published in the *Annals of Coal Mining*, make depressing reading. The big accidents, involving great loss of life, were widely reported, but that was only a part of the picture. Every year there were accidents that caused the death of just one or two people and, according to one witness, although they would be reported to the coroner he would seldom bother to investigate. Originally, the problem had been largely limited to the north-east where gaseous mines were common and the workings were deep. But as deep mining spread to Yorkshire, Lancashire and South Wales, so too did the accidents:

> What with the rapid increase in the number of fiery mines, combined, in too many cases, with insufficient and badly-arranged ventilation, hostility to the use of the safety lamp, lax discipline, ignorance and recklessness of the miners, etc., hundreds of victims were being annually sacrificed to the Moloch of fire-damp; and ever anon the kingdom was startled by the news of some disaster of unusual magnitude, these now becoming of frequent occurrence in districts where they had never been known before.[2]

The *Annals* contain the most detailed survey of accidents involving explosions, many of which were attributed to miners ignoring instructions, and sometimes to safety procedures not being followed. At Barnsley, three girls and a man were killed on 22 February 1842: 'the candles of the three females went out in consequence of choke-damp, but the lad's candle kept alight, being held above his head. This was supposed to have ignited fire-damp and caused the explosion. The steward was censured by the coroner for not examining the pit beforehand.'

The Darley Main Colliery in Barnsley was particularly fiery. On 14 April 1843 an explosion killed one man and injured another. In February 1847 a more serious accident occurred:

> As some miners were firing a shot an alarm of fire was given from a part of the pit 200 yards distant. After two hours incessant exertions, it was found that the fire gained ground, and the pit was so full of smoke that it was impossible to remain. Many succeeded in reaching the pit's mouth in safety though dreadfully exhausted, but six perished.

Just five months later another explosion killed two men, and the coroner's jury added a rider to their verdict that 'through the numerous accidents at this pit there must be some neglect on the part of the managers'. The message does not seem to have got through, and the men continued to work with the naked flames of candles. On 24 January 1849 the worst disaster of them all occurred, announced to the men working at the surface by a dense cloud of smoke and coal dust erupting from the main shaft. By the time the rescuers completed their grim task, they had brought seventy-five dead bodies to the surface. Once again the jury was critical of the proprietors, and strongly recommended that a far better system of ventilation be introduced before work restarted. They also asked the coroner to forward a request to the government to appoint a 'scientific and practical person' to check the ventilation from time to time in all the local collieries and to 'hear any complaint by the workpeople'. There are no further reports of fatal accidents at this pit, but a note was added that there was still a problem with gas igniting from time to time.

A survey carried out in the 1850s of explosions in pits in the north of England listed 163 explosions in a fifteen-year period, of which 136 were said to be caused by the use of naked flames. Yet still nothing was done to remove the danger. There were some attempts to investigate major disasters. After an explosion at Haswell Colliery in Durham, in which ninety-five miners were killed, two eminent scientists were invited to the north-east to carry out an investigation. Professor Charles Lyall was an eminent geologist and Michael Faraday, although he is best known for his pioneering work in establishing the link between magnetism and electricity, had in the early part of his career worked with Sir Humphry Davy in developing the safety lamp. They reached no useful conclusion, but there was at least one moment of humour to lighten the grimness of the task. Faraday had asked the miners how they estimated

Safety was not always taken seriously, even by the miners themselves. One visitor to a Cornish mine found a miner sitting on a barrel of gunpowder, smoking a pipe: 'Had a spark fallen he and his companions must have been blown to bits.'

The safety lamp invented in 1815 is no longer used as lighting in mines but is still used to test for the presence of the explosive gas, methane.

the speed of the air current in the mine, and they offered to show him. They lit a small heap of gunpowder, and timed the passage of the smoke over a given distance. Faraday, being very conscious of the need for safety, asked them where they kept the powder. They told him it was in a bag, kept very tightly tied. 'But where,' said Faraday, 'do you keep the bag?' 'You are sitting on it,' was the reply.[3] They had very kindly given him the only soft seat available, but the eminent scientist was neither grateful nor amused. In the event, the Lyall Faraday report proved to be of little help.

A far more useful report was put together following an explosion at South Shields in 1839. As a result of a public meeting a committee was set up, known as the South Shields Committee, to investigate safety. They were remarkably conscientious, spending three years on their enquiries, during which time they talked to working miners and viewers, consulted scientists and, unusually for a British committee, looked at safety laws in other countries to see if there was anything to be learned from them. Their report, when it appeared in 1843, was a well-argued document, which was summed up by a nineteenth-century historian of the industry as 'far in advance of the standards of the time'.

One of the more startling conclusions reached by the committee was that the Davy lamp could never ensure safety in fiery mines. They recommended two alternatives. The first was the 'improved' Clanny lamp. Dr William Clanny of Sunderland had been working on safety lamps since 1813 and had, in fact, been the inspiration for the Stephenson lamp. He later improved it and in the new version there were gauze cylinders as in the Davy lamp, but the flame was contained within a glass cylinder. Another recommendation was the very similar Mueseler lamp, developed in Belgium, where the mines were notoriously fiery. They recognised, however, that mine safety could never rest just on safety lamps, but ultimately depended on good ventilation. They argued that a prime cause of explosions was the failure to supply sufficient number of shafts in proportion to the length of underground workings. They recommended that it should be made illegal to work any mine unless at least two separate shafts had already been sunk. They looked at other ways of improving matters, such as cutting drifts to release gas, and they were very impressed by a system invented by Goldsworthy Gurney, better known as one of the pioneers of steam vehicles on the road. This used jets of high-pressure steam for ventilation. They wanted all collieries to have scientific instruments for measuring air currents, preferably a little more sophisticated than the method demonstrated to the startled Faraday. There were many other suggestions, including better education for mine officials. One thing, however, was particularly stressed: the need for the establishment of a body of properly qualified government inspectors with the authority to investigate every aspect of the running of any mine.

Over the next few years there was a concerted attempt to persuade the government to set up an effective inspectorate. In 1844, a pamphlet dedicated to Lord Ashley contained a statement by Dr John Murray 'that I utterly despair of any amelioration, unless Parliament interfere'. The argument for inspection was tragically strengthened by yet more explosions throughout the 1840s. Parliament responded with its customary caution by setting up its own inquiry, headed by Sir Henry T. de la Beche, Director of the Geological Society of Great Britain, and Dr Lyon Playfair, Professor of Chemistry at the Royal Manchester Institution, later to become the first professor at the new School of Mines. So these were men well qualified to take a scientific view of the whole question of gaseous mines. They were able to investigate a number of explosions themselves and their report reiterated many of the points made by the South Shields Committee: that although safety lamps were not perfect they should be made compulsory, that bad ventilation was a curse and that an inspectorate would be valuable. Further evidence was offered by Seymour Tremenheere, who had been the commissioner appointed to investigate mines under Lord Ashley's Act. Both reports were cautious, the first stressing that inspection should not be 'over-meddling' and the second suggesting that an inspectorate did not need to have any specific powers of enforcement, but could be useful in making helpful suggestions for improvement.

The Miners' Association petitioned Parliament in 1847 and the result was yet another inquiry – this time into how mining inspection was carried out in the rest of Europe. Eventually, in June 1849, the whole question was put to a select committee of the House of Lords who, after spending a few weeks hearing witnesses, issued a report that was not exactly a clarion call to action. It recommended it as the 'imperative duty' of Parliament:

> to adopt, for the purpose of obtaining such security as is undoubtedly within the reach of precaution, any steps, whether of the nature of inspection or of direct enactment, consistent with the free pursuit of industry and commerce, with the mutual relations in the country between the Government and private enterprise, and with the due recognition of that responsibility imposed upon the owners and managers of mines, which it should be the care of all rather to strengthen than to impair.

It also recommended the establishment of mining schools on the European pattern. The recommendations were so vague that it took another year before legislation could be framed, but on 11 August 1850 'An Act for the Inspection of Coal Mines in Great Britain' was finally approved. It allowed for inspection both above and below ground, and the following year four inspectors were appointed.

The establishment of the inspectorate did nothing to reduce the devastating effects of mine explosions, and with so few inspectors covering such a vast area it could hardly have been expected to. More inspectors were soon appointed, and more legislation followed. But whatever its faults, the 1850 act did one important thing: it established the principle that it was legitimate for government to have a role in private enterprise when the lives of the men and women who worked in the industry were so obviously at risk. It was a start. The act, and the various reports that preceded it, also highlighted another fact of mine life: the big explosions might be dramatic and cause consternation but no one should overlook the smaller incidents that happened all the time, in which workers were injured and killed.

One of the most dangerous times came at the very start and the end of the working day – with the journeys up and down the shafts. In some mines, the workers simply put their legs through loops in a rope to be lowered down a shaft; in others they travelled in the corves, otherwise used for raising coal from the pit. How safe the systems were depended very much on the mechanisms used for raising and lowering. The 1842 Report on Children in Mines gave graphic details of what could go wrong. The dangers of the simplest system, using a hand-operated winch, were touched on in the last chapter and are obvious. The rather more sophisticated version was the horse gin, but its successful use depended on the person in charge and the unfortunate nag that spent its days endlessly trudging round in circles. These creatures were described in the report on Yorkshire mines as 'of the worst description, spavined and blind'. Most gins used some type of balance system, in which an empty corve was lowered as a loaded one was raised. If the motion of the gin was irregular, the corves could bump into each other and collide with the sides of the shaft, and anyone travelling in one of the corves was liable to be thrown out. Even if no one was travelling in a corve, the bumping could dislodge great lumps of coal. Little Enoch Hurst, an 11-year-old boy, was hit by a lump of coal and killed. The official report declared that there was no one to blame, as he could have got out of the way.

The introduction of powerful steam engines to do the job of the horse gin might have been thought to offer a far safer system. In this system, ropes from a horizontal winding drum passed over a pulley at the top of a frame above the shaft. The rope was attached to an iron bar, the 'clath harness', from which the corve was suspended by chains. It too had its dangers, and was actually considered more hazardous than the horse gin. The engine man's only guide to the position of a corve in the shaft during the early years came when a piece of tallow attached to the hauling rope appeared in view, indicating that the corve was 20yd from the top of the gantry. He had to constantly keep his eye fixed, looking for the sign: 'If on the contrary his attention is directed for a moment to another object, you are sent over the pulley with fearful rapidity and killed.' A poor, spavined horse would never have had the power to do this. The other disadvantage of the power of the steam engine was that because it could raise much heavier loads, there was a greater strain on the ropes. These could be as much as 5½in in diameter, but even then could break without warning. William Rayner Wood, who prepared the Yorkshire report, was given a dramatic example of just what could go wrong. He was due to descend on one of his regular inspections underground and was waiting for the corve to reach the surface, when the rope snapped sending the load crashing down the pit. Had it happened a few minutes later he would have been dead. Others were less fortunate. Of the fifty deaths of children reported for the Yorkshire district, thirty-four involved accidents in shafts.

This was one area where real improvements were made throughout the nineteenth century. With the introduction of more and more steam engines for winding, it became a common practice to have the corves running in guides, so that they could not sway and bump around in the shaft. By the 1840s, wire ropes were being introduced to replace hemp, but not without some opposition not from employers complaining about extra costs, but from the men themselves. In spite of assurances that the new ropes were far stronger than the old, there was a deep distrust of novelty. The miners at Jarrow actually petitioned Parliament to have them banned, and the men of Wingate Colliery brought a court case; they lost. Opposition soon faded. It was good news for the owners, because the use of wire ropes brought a huge increase in productivity. Under the new system, far greater loads could be brought up in the cages, and the actual speed of winding at one colliery was found to be double what it had been using ropes.

At the same time as wire ropes were introduced, several safety devices were also tried, both to prevent over winding, pulling the cage up over the pulleys, and to prevent a cage plunging down the shaft. Few of them found practical applications. The first reasonably successful invention was the work of Edward Fourdrinier of Leek in Staffordshire. If a rope broke, strong springs forced iron wedges against the guides to stop the descent. A similar idea was introduced by the American inventor Elisha Otis to be used in elevators. He gave a highly dramatic demonstration of it at an exhibition when he stood on a suspended platform, axe in hand, and hacked through the supports. No one, it seems, was quite so bold or confident to try the same thing with Fourdrinier's device. It did not, however, meet universal success as the wedges were inclined to catch even when the ropes were working normally. Gradually, however, this aspect of mining at least was becoming altogether safer.

Ventilation was a problem that dominated the thinking of almost all the various committees and commissions that sat to ponder the question of safety in mines. Various types of air pump were tried without any great success and the most efficient system was still to have a fire beneath one shaft, topped by a chimney to draw air down another shaft. The dangers of lighting a fire in a coal mine hardly need spelling out. There was a particular problem associated with pillar and stall working. The pillars that had to be left standing while the men were at work in the stalls were much too valuable to be left there forever. Once the miners had left an area, however, the pillar could be removed and replaced by props. But this created a big, open area that could never be properly ventilated. The problem could only be solved by allowing the roof to collapse in to fill the void once the last of the coal had been removed.

European collieries had been working out an alternative method of ventilation for some time, based on the use of large fans powered by steam engines. One of these, consisting of a 22ft-diameter horizontal fan, was brought over from Europe in 1849 and installed at the top of the upshaft at a colliery in South Wales. Over the next few years all kinds of mechanical fans were tried, but the most successful was developed by John Waddle, and manufactured at the engineering works in Llanelli, which had been started by his father. It was a centrifugal fan, in which the casing formed an integral part of the whole machine. Like the other early fans it was powered by steam and set on top of the ventilation shaft; immense fans were built, up to 40ft in diameter.

Blasting was always a potential danger and was regularly used in mines with hard coals, such as those of South Wales. The system had not changed since the early days. It still involved drilling a hole with a metal chisel up to 2ft long that was hammered into the rock. A cartridge of black powder was inserted and a pricker, which should have been made of copper to avoid a premature explosion, created a hole for the fuse, referred to rather alarmingly in one report as a 'squib'. Most of the men in the immediate area scampered off to safety, leaving one unfortunate to light the fuse and get away as fast as he could. The invention of a safety fuse, which could be timed with some accuracy, certainly improved things but it was not always used, largely because of the expense involved. The direct danger to the men responsible for lighting the fuse was only a part of the problem: there was also the far greater danger of the blast setting off fires. A dramatic example of this occurred at one of the collieries in that infamous scene of disasters, Barnsley.

The work at Edmunds' Main Colliery was being pushed forward at great speed, with the men working flat out, winning 50yd of coal a fortnight and receiving substantial bonuses. But the danger signs were plain to see. Small fires started at each blasting and extra men were called in to control them. It was hardly sophisticated fire fighting: the men put out the fire 'by knocking it about with their jackets'.[4] On the Saturday before the accident in December 1862, the men had to work for an hour to bring the flames under control, but the following Monday the blasting continued as usual. This time the fire could not be quenched. The men fought the blaze for an hour and a half to the point where it became hopeless and there was no option but to try to race to safety. Fifty-three men never made it. At the subsequent inquest it was agreed that the working of the pit was 'incautious and unsafe', but no blame was allocated. This was just the sort of thing the inspectorate was supposed to prevent under the provisions of the 1850 act. What had gone wrong? An indignant editorial in the *British Miner* journal demanded that those responsible for the negligence should be made to pay, but it was in the letter pages that the views of the mining community were most clearly expressed:

> The preventive powers of the Act are not and never have been enforced; we want inspection prior to accidents. Sham inquests before mining engineers and viewers, and occasional paltry prosecutions of the small fry in the management afterwards, are ridiculous and ineffective for the prevention of accidents ... The present inspection system is a sham, a farce, and a fallacy; and every intelligent miner knows it.

The miners had complained about the inspectors from the first, pointing out that under the act there was no requirement for those appointed to be professionally qualified, though in practice they were, nor was there any requirement on the part of the owners to listen to the advice they gave. One can hardly blame the inspectors for inefficiency: the inspector for South Wales had so many collieries on his books that, working full time, it took three and a half years to visit them all. The comment about the coroners seemed backed by sound evidence. One of the correspondents to *British Miner* quoted a case where eleven members of a coroner's jury found the colliery overseer guilty of negligence and manslaughter. The twelfth, an overseer like the accused, disagreed. The coroner agreed with the one and ignored the other eleven. It was something of a contrast with the treatment given to a miner who was found with a pipe in his pocket, which he had forgotten to take out before going underground. He was sent to gaol. Miners were not impressed by the workings of the legal system.

Changes were made when the 1850 act was renewed in 1855: new safety regulations were laid down and a scale of punishments introduced. Owners and supervisors guilty of negligence could be fined; miners found guilty of the same crime would be sent to prison. It seemed perfectly logical to the legislators: miners would never be able to afford to pay fines, so they would go to prison anyway. But to the workmen it looked very much like one law for the rich and another for the poor – again. Matters were not improved by what must have seemed another piece of forward-looking legislation, which allowed owners to impose their own safety rules and enforce them through the courts. Interpretation of the law was so wide that one owner insisted on compulsory church attendance as an essential to safety in the mine.

Real progress only came with a new act in 1872, which made it compulsory for all mine managers to pass a government examination of competence; strengthened safety regulations; and introduced some really useful reforms, including compulsory schooling for pit boys.

No reforms could ever make mining completely safe. The environment is intrinsically dangerous. Men working in near darkness, often in very constricted spaces, were always at risk. Injuries were commonplace. Even in the second half of the twentieth century, anyone who went to a mine and ended up in the pithead showers would scarcely see a man in there without some scars on his body. What was notable about the accidents of the nineteenth century was that so many of them were avoidable. To go through the seemingly endless list of disasters, large and small, is deeply depressing, but one very rarely gets an insight into what these meant to the men most directly involved, the miners themselves. The exception is an account of the accident and subsequent rescue operation at Tynewydd Pit in the Rhondda on 11 April 1877.[5]

The men were working very close to the abandoned and flooded number three seam of Cymmer Pit, but with no idea just how close they were. They only found out when they broke through into the older workings and the water came flooding in on them. Most of the men managed to scramble to safety, but fourteen were unable to make it. The four nearest the break were drowned instantly, but two groups of five managed to survive.

As soon as it was known that there were still men trapped underground, the rescue work got under way. The first group was soon found huddled in a pocket of air between the coal and the water. The rescuers at once began hacking at the coal barrier, but no one had allowed for the force of the pent-up air. As soon as the breakthrough came, one man was hurled against the coalface by the violence of the blast and died at the moment of the rescue. The other four were brought safely back to the surface.

Now the hunt was on for the other five, starting on a Thursday. The following day tapping was heard from Thomas Morgan's stall, but from the plans of the mine it became clear that there was a 38yd-wide barrier of coal between the men and their rescuers that could only be approached down roadways that, by now, were filled with water to the roof. The only alternative was to try to reach them through the main flooded part of the mine. Divers volunteered to try to make the 257yd underwater journey to reach the men, but they failed. That only left one alternative. The roadways had to be pumped dry and the rescuers had to hack through the immense coal barrier. Pumping began at once, but even then only four men at a time could work at the coalface in three- to four-hour shifts:

> They rained down blow after blow unremittingly; no halt, no looking back, no word; fiercely, almost savagely, the men worked, and when the shift of three hours had passed only fell back exhausted for fresh men to advance again,

and show that the same grand stimulus inspired them, prompting to the same desperate hardihood and determination.

By the Wednesday, they could clearly hear knocking from the other side, and just when it seemed that the breakthrough was near, gas began seeping into the workings. There was nothing for it but to stop and wait for the engineers to adjust the ventilation to clear away the gas. They were driven away twice, but when the breakthrough was made it almost turned into a disaster, as air began to rush through the hole, and the water began to rise, threatening the whole operation. The hole was rapidly plugged, while the engineers worked out what could be done to ensure there was enough time to get the men out before the space they occupied was completely flooded.

Throughout the whole operation the men worked on relentlessly:

Most wonderful were the endurance and action of the colliers. It was a noticeable feature that, beating against the black face of coal, which any moment might open out and destroy them, they never turned their heads. With blood streaming in the earlier part of the week, from their hands, they yet rained blow after blow and, said a looker on, never turned or paused.

They worked under the double threat of fresh flooding or gas explosions, but no one hesitated. And it was not just the miners who worked – agents, overseer and owners all took their turn, working the pumps and tending the ventilation doors. On the Friday the attempt was made. The face was broken and, as before, a great gale of wind rushed out, but the five men were pulled out before the water engulfed their refuge.

The five men had been entombed for more than a week, with no food at all and only flood water to drink. The ordeal was almost too much for two of the party. One of them, John Thomas, 'kept staring to where the water was and struggling to get away, calling out, "Let me go; I see a hole that I can get away through".' Two of the older miners, George Jenkins and Moses Powell, restrained him and prevented him walking off into the black water and certain death. David Hughes was only a young boy, who later told the author how Jenkins and Powell looked after him: 'The men there saved my life. They nursed me all the time. I was kept warm by sleeping in their laps.' At the end of his ordeal he had only one wish to make – that he would never again be sent down a mine.

In the long list of mining accidents in the nineteenth century, Tynewydd would go down as comparatively minor, with only a few lives lost. Nor was the rescue effort in any way exceptional: it was what miners did at any pit when disaster struck. It is only special because it was one of the very few for which we have a first-hand account. We can picture for ourselves the desperate work of the rescuers and the appalling conditions that the trapped miners endured, never being certain when, or even if, help would arrive. However tragic such events were, they tied the men together in tight bands of mutual dependence; every rescuer knew he could be the next victim and each victim might be called on in his turn to become the rescuer. It was not just the men directly concerned who were involved either but the whole community. Wives and families gathered at the pithead, waiting for news, comforting and encouraging each other. They were locked in their own little tragedy, largely ignored by the outside world. It was this sense of mutual dependence, of needing to rely wholly on each other with little hope of outside help, that gave these communities their coherence. It was this that enabled them to fight together against a hostile environment, and to join together to fight that other battle for a better way of life.

CHAPTER NINE

THE FIGHT FOR UNITY

In the early 1840s, the various miners' unions were in a state of considerable disarray. The owners considered them broken and felt they could do more or less as they wished. In the coalfields of the north-east, wages were reduced time after time and there seemed to be no spirit left to attempt any resistance. But a new movement did begin to stir and the same period saw the formation of the Miners' Association of Great Britain and Ireland. It grew slowly at first but a document was drawn up and sent to the coal owners. It set out what the association hoped to achieve:

> Our object in forming the above association is to better our condition, and we beg to apprize you that we would rather by far that could be done by an amicable adjustment of all differences, than by having recourse to a strike, which we feel inclined to believe is equally disadvantageous to you as to us, and the inevitable result of which would be to engender feelings of such a kind as ought not to exist between master and servant. We intend to lay before you the following specific and simple plan, viz., that each colliery owner shall be furnished with a copy of such prices as shall be thought necessary and reasonable, and in which it is intended to go on the principle of making the cost price, as far as labour is concerned, equally or nearly so in every colliery in the trade, and to such uniformity of cost price the masters to add what they deem a proper and reasonable return for their capital. It being our firm and decided opinion that, as we risk ourselves and you your money to dig from the bowels of the earth a commodity on which it may truly be said the existence of Great Britain as a nation depends, it is not too much to request that the price of that article shall be such as to give ample remuneration to both the labour and capital employed.

The document continued by referring to the fact that the different mines were indulging in a cost war, each trying to undercut the other, and that the miners were seeing the results in lower wages. It was never likely that the owners were going to agree to the main proposal, but it would at least have been a step forward if they had agreed to meet the union representatives. The union leaders waited for the response. What they did not expect was no response of any sort at all from any mine owner. They were just completely ignored. A second petition was presented, setting out the grievances of the Durham and Northumberland miners, particularly the system of payment by measure, which was open to abuse. They were met by the same absolute silence.

In March 1844 a general meeting was held in Glasgow, attended by delegates representing some 70,000 men from different parts of Britain. High on the agenda was the condition of the miners of Northumberland and Durham, and there were calls for a general strike of all the collieries to support their claims. This was more than a little ambitious. There was no real sense of unity between the various districts and one delegate from Lancashire wryly noted that if the resolution was passed the next meeting would be to attend the union's funeral. The motion was narrowly rejected. All the delegates did, however, hear at first hand just what the men of the north-east were suffering. Many of the complaints were the familiar ones. One man reported how they were paid 4½d a tub but if the tub contained small coal they were fined 6d, though the small coal sold at 6s per chauldron. A hewer at Tyne Main described how 'the laid out is something fearful here'. The 'laid out' were tubs that were not considered satisfactory. He was supposed to chalk up the different types of coal but was uncertain how this should be done, and the overseer could not advise him. In the event, he sent up nine corves in the day and eight were laid out:

> This man worked for 1s 6d, and there was 2s kept off him; so that he laboured all day for nothing, and had to pay the masters 6d for allowing him to do so. Kind Heaven look down upon us, and guide us the way to get clear of this repression, for the miner's cup is about full. No human being can bear the treatment which is daily inflicted upon us![1]

Once again the association wrote to the owners, setting out their grievances and requesting a meeting, and once again they were met only by silence. The next step was inevitable: 'The men felt insulted by the contempt and, goaded by the insolence of their employers, they resolved that the men of the two counties should cease working until their differences were adjusted.' The strike began on 5 April 1844 with a mass meeting attended by the union leaders and an estimated 35–40,000 miners. It was a cheerfully optimistic occasion; banners flew, flags were waved and bands played; a day to remember. The mood was short lived; though soon the two counties would be overwhelmed by a tide of misery and deprivation.

Agents were sent to the coalfields of Britain to recruit replacement labour and soon there were enough to begin limited work at the pits. The owners needed to find homes for the newly arrived miners and their families, which was not a problem: they simply evicted the strikers and their families. Camps sprang up around the colliery villages, such as at Seaton Delaval:

> In one lane, between Seghill and the Seaton Delaval avenue, a complete village was built, chests of drawers, desk beds, &c., forming the walls of their new dwelling; and the top covered with canvas, or bedclothes, as the case might be. Here and there, fiddles might be heard; whilst the men grouped together, smoking, singing, or chatting about the great battle, but never wavering in their confidence, or in their determination to fight out the battle to the bitter end.

Large numbers of men were recruited from Wales and other regions, usually arriving with either a police or military escort. The local men tried to persuade them to return home and several said they would do so, but they didn't have the funds to pay for the return journey. The locals had a whip round to pay the fares and some of the Welsh did leave, but a minority took the money and then went to work anyway. Their action was neither forgiven nor forgotten. There were strenuous attempts by the union to act as if there was indeed some sense of unity between the different districts. A delegation was sent up to Scotland to put their case to the local men. This was regarded as deeply suspicious by the local authorities. On the day of the public meeting, the magistrates sat in readiness and the military were put on standby and surrounded the hall. It was packed out anyway and everything passed peacefully. Some authorities took the matter even further. A similar meeting at Bedworth in Warwickshire was declared illegal by Lord Lifford, the yeomanry was called up and the police instructed to prevent anyone from speaking. In the event a very large crowd appeared, ready to hear what the men from the north-east had to say, and the constables wisely decided that an orderly retreat was their best option. Again, the meeting was completely peaceful and Lord Lifford was unable to find any law that had actually been broken.

The union was still hopeful that something could be achieved by negotiation, and they asked Lord Londonderry, one of the most powerful landowners in the north-east, to act as a mediator. He refused the invitation, but he did take action. Seaham Harbour was part of his estate and he issued an order to all the local shopkeepers to stop offering credit to any of the miners or their families. Meanwhile, the agents were scouring the country to recruit miners. A group arrived from Staffordshire who had been told the strike was over and they were wanted for a new mine that was being opened up. When they learned the truth they turned round and went home. Men were even brought from as far away as Cornwall, with extravagant promises of high wages. They were signed up, but within a fortnight they found that they were being offered 4d a tub, and they needed to fill twenty-four a day to make the 4s they had been promised. The most experienced coal miner could never have managed that, and none of these men had any experience in hewing coal, a very different type of work from that of the tin and copper mines. They were, in fact, rarely managing more than four tubs a day. All but four of the thirty-two men absconded. What followed was pure farce.

A £50 reward was offered for the capture of the absconders. Four were captured and put under police guard, but managed to escape again. There was a great hue and cry, joined in by colliery officials. One astute officer claimed to have found 'a Cornish footprint' and another discovered one of the absconders, only to find himself overpowered and beaten with his own truncheon. Some of the Cornish were arrested by police in North Shields and a steamer was hired to take a party of special constables to bring them back to Alnmouth. There they were brought up before the magistrates and, to the great delight of the strikers, they were all acquitted.

This was a small victory for the strikers, but after fifteen weeks most of the collieries were at work, even if production was far lower than it had been. Rumours began to circulate that some men were drifting back to work, which was not surprising given that so many were still living in the makeshift camps, where poverty and exposure to the elements was taking its inevitable toll through disease and malnutrition. Families pawned everything they had, including their wedding rings, to pay for food and even shared what little they had with those who had nothing left to pawn. One group of musicians took their instruments out to play in the street; the 'leader' of the band was arrested, charged with begging and sentenced to fourteen days. On 30 July another meeting of the strikers was held, but fewer turned up this time. There were still banners with brave slogans:

> Stand firm to your Union,
> Brave men of the mine,
> And we'll conquer the tyrants
> Of Tees, Wear, and Tyne.

However, many must have felt that they were further from victory than ever. Their greatest bitterness was reserved for the men who had taken their places. Trouble erupted at Seaton Delaval, where the Welsh now occupied the cottages of Double Row. It started with a small affray, when two Welsh miners went out in the evening to fetch beer from the local pub. They were met by local men, a fight broke out and soon others joined in. It became a vicious battle in which any weapon that came to hand, from pick handle to fence post, was used and a number were seriously injured. It solved nothing other than to relieve frustrations: the pits remained open.

The situation became desperate. Families even applied to enter the local workhouses but were turned away from the door. With no funds and nothing left to pawn or sell, the strikers had no option: by the end of August the strike was over. Not everyone had jobs to return to and those who did go back found themselves having to work alongside the Welsh and others who had chosen to stay. The atmosphere was as bad as it could be, and the local men did all they could to make life hard for the intruders; they even took tags off the tubs they had filled to stop them being paid. The hardest hit among the local men were those who had been identified as strike leaders, who found themselves blacklisted wherever they applied for work.

The strike of 1844 was a massive defeat for the new association. Not only had they failed to win better conditions for the men they represented, but those who went back had found themselves actually worse off than they had been before they came out on strike. It was also clear that although the association might claim to represent all sections of the mining community throughout the country, it was powerless to prevent the use of blackleg labour. The dream of a united body, capable of representing all the collieries and powerful enough to stand up to the owners, seemed as far from realisation as ever.

One of the stated aims of the association was to see an effective system of government inspection and they could at least claim some success here, with the passing of the 1850 act, for which they had lobbied Parliament. How much influence their views had on the actual decisions is difficult to tell, but at least something had happened. In the event, they were to be deeply disappointed at the way in which the system was put to work.

The man in charge was Hugh Seymour Tremenheere, though his background gave little indication of why he would be the best man for the job. He was educated at Winchester and Oxford, where his main claim to distinction was translating Pindar's odes from Latin into English. He had, however, spent some time investigating conditions in Cornish mines, where he found that the average age of death of the miners was 43 compared with 54 for the rest of the population, due to what was then called 'miners' consumption'. He recommended better ventilation for the mines and the introduction of pithead baths, and he was particularly keen on education, especially religious education. He was very clear that education was not meant to help the miners improve their lot in life. It was to provide 'moral and spiritual improvement: not to raise the individual from his own sphere but to enable him to do his duty in that in which he belongs'.[2] When he was appointed he wrote to the colliery owners to explain the workings of the act. On his tours of inspection if he found the act had been broken he issued warnings, but did not always enforce the law. On at least one occasion where he found very young boys working in narrow seams he refused to act, on the grounds that the owners had convinced him that the work of the young lads was necessary if the mine was to make a profit. Martin Jude, one of the leading officials of the Miners' Association, was scornful of a visit he observed. Tremenheere, he declared, 'did not come to get a correct report'. He was 'taken to the colliery office and courteously assisted to the station by officials; but the men were not in a single instance consulted'. Tremenheere himself rebutted these charges, stating that he was 'always seeking interviews with the most violent of the agitators'. Whichever view is correct, it is very clear that the miners themselves had little or no faith in the work of the inspectorate. If anything was going to be done to improve their working condition, they would probably have to do it themselves.

One lesson the miners had learned was that as the owners were always likely to try to use the law for their own ends as far as possible, they needed a qualified lawyer of their own to put forward their side of the story. William Prowting Roberts was employed first by the Northumberland and Durham Miners' Union, then by the Association and, when they ran out of funds following the 1844 strike, by the newly formed Lancashire Miners' Union. He became popularly known as the 'miners' attorney-general' and proved to be a powerful advocate. In a letter to another trade union, he described the difficulty of dealing with magistrates. He thought they were on the whole honest, but 'all their tendencies and circumstances are against you':

> They listen to your opponents, not only often, but cheerfully – so they know more fully the case against you than in your favour. To you they listen too – but in a sort of temper of 'Prisoner at the Bar, you are entitled to make any statement you think fit, and the Court is bound to hear you; but mind, whatever you say,' etc. In the one case you observe the hearty smile of goodwill; in the other the derisive sneer, though sometimes with a ghastly sort of kindness in it.[3]

He went on to point out the dual standards of the day: a group decision by the owners not to employ a 'troublesome fellow' was perfectly legal, 'reverse the case, however, and it immediately becomes a formidable conspiracy, which must be put down by the strong arm of the law, etc.'. But he ended up on a more optimistic note. The oppression he fought against, 'which after all is merely a more genteel and cowardly form of thieving', could be beaten by using all the powers of the court. He did achieve successes, notably in leading the fight to defeat a bill that would have given magistrates sitting alone the right to hear a complaint from an employer and then issue a warrant for the workman's arrest and imprison him. Among the enthusiastic champions of Roberts' work was a young German from a wealthy family of textile manufacturers, who had been sent to Manchester to study English production methods. His name was Frederick Engels. He described a whole string of cases in which the magistrates had condemned men to prison, only for Roberts to get the verdict overturned by a higher court. A typical case was one in which three men from Staffordshire, who had refused to work in a place that was threatening to cave in, were brought to court for breaking their contract. Although they proved totally justified, as that section did indeed cave in before the case was heard, they were still sentenced by the magistrates, but later had their conviction overturned by a judge. Engels went on to list other cases:

One after another Roberts brought the disreputable mine owners before the courts, and compelled the reluctant Justices of the Peace to condemn them; such dread of this 'lightning Attorney General' who seemed to be everywhere at once spread among them, that at Belper, for instance, upon Roberts' arrival, a truck firm published the following notice:

NOTICE!
Pentrich Coal-Mine

The Messrs. Haslam think it necessary, in order to prevent all mistakes, to announce that all persons employed in their colliery will receive their wages wholly in cash, and may expend them when and as they choose to do. If they purchase goods in the shops of Messrs. Haslam they will receive them as heretofore at wholesale prices, but they are not expected to make their purchases there, and work and wages will be continued as usual whether purchases are made in these shops or elsewhere.[4]

These were small victories to celebrate, but there was a limit to what Roberts could achieve and law never came cheap. Roberts himself was paid £1,000 a year and all cases incurred extra costs. Once Roberts had moved away from a district, the legal pigeons that had fluttered nervously in their dovecotes soon resumed their more familiar hawkish appearance. By the late 1840s all the different unions were struggling financially and the association, which had never recovered from the 1844 strike, finally collapsed. The ideal of a national union seemed further away than ever from realisation.

There followed a period of spasmodic growth, with a number of areas forming purely local unions based on individual collieries that slowly began to amalgamate to form county unions. There was also a great debate over whether or not a national union was really desirable and, if it was, what form it should take and what policies it should pursue. One possible road to go down was opened up with the formation of the British Miners' Benefit Society in 1862. This was very different from earlier unions. Its leading figures were mainly aristocratic and they set out their aims: to promote scientific research into mine safety; to educate the miners; and to provide an insurance fund for deaths and injuries. These were all very sound aims. Lord Raynor toured the colliery districts to explain the plans to the miners and must have been delighted by the reception he received, with miners pouring into the public meetings. At Wigan it was estimated that 15,000 turned out to hear him. If it was successful in attracting the support of the majority of the mining community, the society would also have been a powerful body for lobbying Parliament. But it soon faced serious opposition.

First to come out against it was the trade magazine *The Colliery Guardian*, in which an editorial spoke sarcastically of its leaders as 'philanthropic lawyers, patrician noblemen and such-like ornaments of society'.[5] They were, it was said, interfering in matters of which they had no practical knowledge and would do more harm than good. The owners had good reason to oppose the society. Many ran their own benefit societies, which were very useful for preventing men leaving to find better pay and conditions at other pits. Some even threatened to sack any workers who joined the national scheme. Similarly, many of the local unions also ran benefit schemes, and they preferred to continue doing so and had no inclination to hand over to his lordship and his friends. Attacked from two directions, the Benefit Society never really got under way. However, it had raised the interesting question: was there a better way to deal with grievances than by strikes and confrontations?

One man who had backed the society was John Towers, the editor of the *British Miner*, who was a keen supporter of the idea of working through persuasion rather than force. It was not a very popular view with many of his readership, however, and he began to worry about falling circulation figures. He responded by making the paper an open forum for miners to express their views, which they did with enthusiasm. It was in the paper's letter columns that a new movement towards forming a national union was begun. The first initiative came from Lancashire, where there was a successful union and they suggested forming an alliance with Scotland. Others soon joined in and a delegates' conference was held in Leeds on 9 November 1863.

At once their differences began to appear. The men of the old 1840 union wanted to continue with an aggressive policy of fighting the owners at every stage. But the opposing idea of negotiation and pressing for legislation had a powerful advocate, Alexander MacDonald. He was very much representative of a new generation of mine leaders. He had started down the pit at the age of 8 in 1829 but, against all the odds, had managed to educate himself to such an extent that he was able to enter Glasgow University, working down the mine in vacations to pay his way. He was a pugnacious, loud and rumbustious character who knew where he wanted to go and was none too gentle with anyone who got in his way. He was totally convinced that the old ways simply had not worked, which was certainly true, and that more was to be gained by non-confrontational negotiation, which had not necessarily proved that much

The nineteenth-century journal *British Miner*, edited by John Towers, became a forum for miners from all over Britain to give their opinions and share their grievances. It was through the letter columns that a new miners' union was formed in the 1860s.

more effective in the past. What he could and did argue was that, not only had strikes not provided the hoped-for results, but they had brought extra misery and poverty on the strikers. His view won the day and he became the leader of the new union. It was not long before a test case appeared, in which the different attitudes would be highlighted.

The workers at Blaina Colliery in South Wales were complaining that they were paid by the weight of coal they produced, but they had to hew 30–35cwt just to get paid for 1 ton. They asked for an independent checkweighman to be appointed, as was allowed under legislation of 1860. The company refused so a protest strike was called, at which point the company had nine men arrested and charged with breach of contract under the Master and Servants Act. The defence was led by Roberts, who found an all too typical situation, namely a magistrate who was a friend and neighbour of the mine owner. He promptly went on the attack. He brought a counter-accusation against the company, calling as a witness a young, orphaned girl who was being required to pay off debts owed to the company shop by her dead parents. She had been working for months and had only been paid in goods, not a penny of cash. This was clearly illegal under the Truck Act and a great embarrassment to the company. A verdict was quickly reached, in which the men were given sentences of one day each, which meant that as they had already been held for four days they walked free. In South Wales it was considered a great victory, but MacDonald was horrified. This was the sort of aggressive campaigning he was fighting against.

It was clear that the new union had already split into two factions: one led by MacDonald; the other by the old school, headed by Roberts and Towers. It was as much about a clash of personalities as it was about tactics, and the result was almost inevitable. A new, breakaway union was formed, the Practical Miners' Association. Ultimately, the split did nothing to encourage local mines to give up their autonomy. The Practical Miners supported a strike in Staffordshire, which achieved nothing and effectively used up all the funds. It was a disaster from which the union never recovered and it simply slid into oblivion.

One thing that did emerge from the struggle was the growing importance of the *British Miner* as a forum for ordinary miners to make their voices heard. It provides a unique insight into the miners' lives and what they themselves thought about their conditions. Two topics emerged time and time again: butties and truck. The butties were overseers, many of whom made a very useful extra income by running drink clubs. It was known as the 'drinking system' and at its most blatant it required the miners to pay a portion of their wages into the clubs, in return for which they got a ration of beer. The butties decided how much beer each man should have and the price he would pay for it, and anyone who objected was likely to find himself getting all the worst and most badly paid jobs in the mine, assuming the butty didn't actually arrange for him to get sacked. The men who got a reasonable amount of beer at a fair price still had to take it under coercion; those who failed to get a reasonable amount at a reasonable price were simply being robbed.

Truck was regularly denounced and the men were not afraid of naming names. In the issue for 31 January 1863, James Merry, a mine owner and Member of Parliament, was referred to as 'the great truck king of Scotland'. He employed more than a thousand workers in his mines and ironworks and paid them monthly in arrears. As the men never quite caught up with the payments, they were given advances, but they came with a proviso: the money could only be used in the company shop. The company made regular deductions from wages: 2*d* a week for the doctor; 2*d* for school; money for the blacksmith to sharpen their tools; and money for the rent on company houses. But the one that seemed to rankle most was the stoppage of 4*s* a month for coal, as they were actually paying more for getting coal direct from the pithead than they would have done if they had bought it from any local coal merchant.

The other common grievances concerned the vexed question of payment by weight and, whatever official policy might have been on strikes, there were occasions when the men saw no other options as being open to them. At Methley Junction Colliery the payment system was changed with just two days' notice. Previously it had been very simple: they got 1*s* 1*d* per ton. Now the men were required to sort the coal into big and small pieces, which was 'putting a deal of extra work on the men'. The rate for large coal went up by 2*d* to 1*s* 3*d* a ton, but for small coal it went down by 4*d*. No allowance was made for the extra time spent on sorting. The colliery paid on 5cwt measures, which inevitably were rounded down and never up, so that if a man brought up 19cwt he was only paid for 15cwt. The writer described an occasion when, due to rounding down, they had lost out on payment for over 10 tons on a single day. A deputation went to ask for better rates, similar to those paid in nearby collieries: 'All the answer we got was that he (our master) made his own laws and we must go by them, he had nothing to do with our neighbours.' The letter ended: 'It has been rumoured that we are sullen and stupid, and will not go and talk with our masters; but we have been and have told you of our reception.' Similar stories were told from other pits. Sometimes there were variations on the theme. There were collieries where men were paid by the tub; it would be filled at the coalface, but in its long journey to the surface, the coal would shake down, some would fall off and at the end it was declared as not full and no payment was made at all. In others, the coal was passed over screens to separate the small from the large; the screens were known by the owners as 'Billy Fair Play' – the men called them 'Billy One Side'. There were some victories to record. In Fife, the owners were successfully prosecuted for using a steelyard that under-measured by 12.5 per cent.

Towers had never intended the journal to represent the voice of militancy, but that is what it had become thanks to the readers. The mine owners hated

Cornish miners at Dolcoath in the 1890s. Because the tin and copper mines were not gaseous they could use candles for light. (Royal Institution of Cornwall)

Scottish miners in the sort of costumes they wore in medieval times, with their simple implements and tools, carved on a tombstone. (National Mining Museum, Scotland)

An illustration from Agricola's *De Re Metallica* (1556) showing a variety of devices used for hauling materials up and down the mineshafts. (Author's collection)

An illustration from the 1842
Parliamentary Report on Children in
Mines, showing a boy and girl being
lowered down a shaft using a simple
winch. (Author's collection)

Scottish women carrying coal up
near-vertical ladders in the late
eighteenth century. (Author's collection)

The penitent: the miner, covered in protective sacking, is lighting a pocket of methane gas using a candle at the end of a long stick. (Author's collection)

In the twentieth century, hydraulic roof supports replaced the old wooden pit props at most collieries. (National Mining Museum, Scotland)

Miners were usually paid as a gang; here they are sharing out the money on payday. (National Mining Museum, Scotland)

The mine rescue team in Fife in the early twentieth century. (National Mining Museum, Scotland)

Miners' tenements in Smeaton, Fife, photographed in the early twentieth century. (National Mining Museum, Scotland)

Few miners had the luxury of a bath in their homes so they were forced to wash in the kitchen. (National Mining Museum, Scotland)

The children of miners evicted from their homes by the owners of Kinsley Colliery, West Yorkshire, being fed in the street in 1905. (National Mining Museum, Scotland)

The organisers of a soup kitchen for striking miners in the 1920s at Newton, Midlothian. (National Mining Museum, Scotland)

Using a mechanical coal cutter at Canderigg Colliery. (National Mining Museum, Scotland)

Five miners were killed when this roof collapsed at Seafield Colliery in 1970. (National Mining Museum, Scotland)

Police dragging away a picketing miner during the 1984 strike, at what became known as the Battle of Orgreave. (Royal Institution of Cornwall)

it, and a news item of May 1863 described the eviction of two miners and their families for the crime of selling it. One of them had a child that was seriously ill and they feared for its life, since none of their neighbours dared take the child for fear of being turned out on the street themselves.

The paper did not just consist of stories of struggle and deprivation. There was a weekly fiction serial and a strong emphasis on self-improvement. A letter from Dudley described how the miners had a reading room and used it for meetings. One occasion, described as 'a social tea party', offered an interesting agenda for discussion: 'the principles of capital punishment and popular amusement &c.' The unlikely connection was not clarified. Such stories reflected the growing influence of Methodism in many areas.

There was another paper giving news and views on the world of coal mining. *The Colliery Guardian* was the voice of mining engineers and management. It often contained the same stories, but put a very different slant on them. In the issue for 27 March 1858, for example, there was an account of a strike in Yorkshire against a reduction in wages of 15 per cent. The response was familiar: examples were made as a warning to others. One miner, William Moorhead, was convicted of leaving work without giving one month's notice and was sentenced to one month's hard labour, a result the paper greeted with considerable satisfaction. Miners were occasionally given a say, but only when they toed the management line. An opponent of the Yorkshire strike was found who wrote that 'if we act with care and caution, and go about our business in a proper and legitimate way, we may persuade our masters not to be so hard on us'. Even the writer did not seem too hopeful that this technique would work.

Following the schism in the union movement, a new attempt was made to create a national body, with the formation of the Amalgamated Association of Miners under a new leader, Thomas Halliday, who had previously worked with MacDonald but had become disillusioned with the old movement. It was based on a loose federation of local unions, but with a central controlling body that would manage funds to support strike action anywhere in the country. It soon gained strength in South Wales, and it was here that it was put to its severest test.

The story was a familiar one, but made rather more complex than usual by the fact that although the owners tended to act together, they were actually made up of three distinct groups: the Steam Coal Colliers' Association, the Ironmasters' Association and the rest, collectively known as 'non-Associates'. It all started with the imposition of a 10 per cent cut in wages, but the Ironmasters broke rank and changed it to just 5 per cent, and the rest reluctantly fell into line. Having won a concession without having to do anything, in May 1871 the men put in a request for a 10 per cent rise and again it was the Ironmasters who responded by offering 5 per cent, which was accepted. This time, however, the Steam Coal Colliers' Association decided to hold out against any increase and the men went on strike.

Earlier descriptions of strikes have generally been taken from accounts either written by union officials or authors sympathetic to the miners' cause. This time, however, the most complete account comes from the leader of the owners' association, Alexander Dalziel.[6] He makes it very clear that the argument was not just about wages; the new union was a threat that had to be resisted. The formal resolution of the owners' meeting stated: 'unless the proprietors resist to the utmost any demands of the men for an advance of wages at the present time, it will be impossible to manage them.' Dalziel added his own comment that this was more important than 'the mere monetary consideration'. The owners called on a notorious strike-breaker, Paul Roper, to recruit 500 men and boys, and they agreed to pay him 6s for every one he recruited; even Dalziel seemed to find him an unsavoury character. The chief recruitment areas were Cornwall and Staffordshire. The new unity was put to the test and the Staffordshire miners sent an encouraging letter: 'Respectable men will not come into South Wales; nothing but the very scum of the country will put their feet upon your soil … stand firm. *Don't be downhearted.* We will stop all we possibly can.' They were as good as their word, and posters and handbills were widely distributed throughout the county:

MINERS!

The bill of Paul Roper, stating that 500 men are wanted is ALL DECEPTION. They are only wanted for a few weeks in South Wales, where 9000 men have been on strike for eight weeks. Miners of South Staffordshire! Stay at home, and the South Wales case is sure.

Inevitably some men were lured to Wales, but they could only get to work under police escort, and the owners were so worried about the support that the miners had among the community as a whole, that instead of making plans to call in the local militia, they disarmed them. In time the Welsh managed to cajole or persuade the strike-breakers to go home, at which point the owners had little choice left but to go to arbitration. The result was the award of 2.5 per cent immediately, with a further rise to come, and agreement that in future they would match rates already agreed by the ironmasters. It seemed a famous victory, but there was little time to celebrate.

At the end of 1872, the owners struck back by announcing a 10 per cent cut. The miners proposed arbitration, but the owners refused. They banked on the fact that funds would be low after the last strike and 60,000 men were locked out. Eventually, the demoralised men agreed to a compromise with a lower cut. The owners sensed they now had the union on the run, and the men were scarcely back at work before they announced yet another cut. Halliday came down from Lancashire to help organise a strike and brought a £1,000 strike fund from the Durham miners. The owners responded by going on the attack again. They announced the 10 per cent cut was no longer on offer; the new cut was 15 per cent. Many men were thrown into a panic, worrying where the cuts would end and a slow drift back to work began. Eventually, the men accepted a 2.5 per cent reduction. That was not the important thing for the owners, for they had achieved their main objective. They had broken the power of the union in Wales. The defeat also spelled the end of the Amalgamated Miners' Association. To many it seemed the end of all hopes for a united organisation to represent all the miners in Britain.

CHAPTER TEN

ABOVE AND BELOW GROUND

The second half of the nineteenth century could be thought of as the 'Age of Steam'. Until 1800, thanks to the Boulton and Watt patent, all development of steam engines other than by engineers approved by the company – and virtually no one was approved – was prevented. Within the company itself development had also been limited by Watt's insistence on only using low-pressure steam. When the patent ended in 1800 engineers, notably Richard Trevithick, began using high-pressure steam. In the first place this meant that engines could be much more compact, but the real change came when he placed the steam engine on wheels and set it off to haul a train of trucks down a railed track. It was not an immediate success, but after a slow start the steam locomotive came into its own. The opening of the world's first intercity railway in 1830, linking Liverpool and Manchester, marked the start of a period of rapid development that soon saw the whole country covered by an intricate rail network. Steamships were soon crossing the oceans of the world and the steam engine had become the source of power of choice for factories and mills. And they all needed coal to fire their boilers. A new use for coal was also found. Heated in retorts, it turned into coke and gave off a gas that was used for lighting streets and the wealthier homes. The nineteenth century also saw a huge movement of the population, from the countryside to towns. Now, instead of being able to gather wood and other fuels, they had to rely on coal for heating. It all added up to an unprecedented demand for more and more coal, a demand largely created by new technologies. One would expect to find the same technological breakthroughs in mining that were transforming the rest of the country's industries.

Above ground the scene at mines in the second half of the nineteenth century certainly looked very different from that of a hundred years earlier, showing that here too the age of steam was having an impact. Most mines had at least one engine house holding a powerful pumping engine, perhaps a second to wind men and materials up and down the shaft and, in mineral mines, a third to power the stamps that were used to crush the ore. Moving the raw material away from the mine was made more efficient, with many mines having their own railway systems linked in to the rapidly developing national network. A very good example of a reconstruction of a nineteenth-century pithead scene can be seen at the open-air museum at Beamish in County Durham.

Here there is a vertical steam winding engine of a type invented in 1800 by Phineas Crowther of Newcastle and much favoured in the north-east. The winding drum was set directly above the steam cylinder. The tubs were loaded into a cage then drawn up to a staging at the top of the shaft, where they could be wheeled across to a tipping device, which sent the coal shooting down to the screens below. The small stuff fell through, and the rest was shot down into waiting trucks on the railway under the screens. A locomotive would then have hauled a full train off to the Wear or the Tyne for shipment. It was all very efficient, and manual labour was largely limited to sorting out the rubbish, such as stones, from the coal as it passed over the screens. It was a very different story below ground.

There were attempts to improve haulage below ground using steam power, including a system devised by George Stephenson, but none were really effective. Throughout the nineteenth century, virtually the only system that didn't rely on human effort to move tubs below ground was one using pit ponies, and they could only be used in the broad, main roadways of the mine. Attempts to mechanise coal cutting were not much more successful. As early as 1768 trials had been made of a mechanical pick, which attempted to reproduce the swinging action of a pickaxe through a system of gear and levers, powered by two men working a crank. It was soon apparent that two men using conventional pickaxes were more efficient than the same two men working the machine. It was never heard of again, and further progress was hampered by the same old problem: the lack of a power source that could be safely used in mines. The steam engine was not the answer – no one wanted to be firing a boiler underground – and attempts to work machinery through linkages to a surface engine resulted in such loss of power as to be useless.

Development depended on finding a different power source that could be used safely in mines. The answer was not exactly new. The German physicist Otto van Geuricke had been experimenting with compressed air as early as the seventeenth century, but the first patent for a practical compressor was not taken out until 1829. In 1861 Thomas Harrison, an engineer from Durham, took out a patent for a coal cutter using a compressed air turbine for power. In its first stage of development the cutter consisted either of a disc with a serrated edge, like a circular saw, or one with cutters mounted on the rim. In later versions it had a hollow box into which cutting devices could be fixed, one of which was very similar to the familiar chain-saw of today. The designs were sound, but the problem was still power loss, because the compressor was too far away from the cutter. Machines of a similar, basic design only really became practical at the beginning of the twentieth century, when the compressors were powered by electricity.

One important mining operation was improved. Blasting required holes as deep as 3–4ft and 1–3in in diameter to be drilled. There were various attempts to produce powered drills, including a rotary steam drill invented by Richard Trevithick to help in quarrying stone for the construction of the Plymouth breakwater in 1813, but none led to immediate developments. Success finally came with the appearance of the Ingersoll compressed air rock drill in 1871. After quite a short development period, the new type of drill was reduced in size so that it could be operated by just one man. This was a very attractive proposition to the mine owners, as work was not only much faster but the old system of drilling by hand had required two men, one to hold the chisel and the other to strike it with a hammer.

The arrival of the pneumatic drill proved to be a curse rather than a boon to Cornish miners. Mortality had always been higher in the mining population than in the rest of Cornwall, but early death had generally meant men dying in their forties. A Parliamentary commission reporting on the health of Cornish miners in 1902 found that by the 1890s things had got a lot worse:

> During the last few years, however, there has been an enormous increase in the death-rate from lung diseases, particularly among younger men from about twenty-five – forty-five with the result that the total death-rate at all ages is now far greater than at any previous period during the last fifty years. Between the ages of twenty-five and forty-five the death-rate from lung diseases among miners living in Cornwall has recently been from eight to ten times the corresponding death-rate among coal miners and ironstone-miners.

The problem was attributed to the dust thrown up by the machine drill, boring into solid rock. It may have made the job easier and the work more efficient, but there was a high price to pay. There was potentially another factor involved in the increase in lung disease. Men working overseas in very bad conditions were returning to Cornwall with their health already ruined.

In effect, no machines apart from the pneumatic drills made any real difference to life underground. In other respects, life went on much as before. Most of the hewing continued as it always had, with the men undercutting the coal in soft seams by hand, then levering it down with pickaxe or crowbar. Hard coal had to be blasted, the one part of the work that had been changed by technology, but after blasting the job of breaking up the coal and loading it into the tubs still depended on muscle power. The basic conditions did not change either: men still had to work in cramped conditions, especially in narrow seams where, in some cases, it was not even possible to kneel to work but instead they had to lie on their sides. Most people have probably wielded a pickaxe at some time, or have at least seen one used. The effort is mainly in raising the pick, after which gravity helps to give power to the actual down stroke. When you are lying on your side, everything is down to unaided muscle power. For the working miner, life in the second part of the nineteenth century was very little different from how it had been a hundred years before.

Life below ground had changed very little, as had life in the colliery villages and towns. One of the more interesting accounts comes from an anonymous lady who married a mine owner in the Midlands.[1] Her friends reacted with the sort of shock and horror that might have been expected if she had announced she was off to explore the Amazon by canoe. When she arrived she was more than a little apprehensive, but found the miners not to be the savages she had been led to expect, though her first sight of them was not very encouraging. She met men as they 'loosed out' at the end of a shift and found them to be 'a rough set, a colony of sweeps'. She was surprised to find that for the journey to and from the pit, they almost all wore wildly impractical white flannel jackets. She also found it strange at first that there were so many men lounging around after three in the afternoon, until she discovered they were the hewers, who had started work at three in the morning. She reported a certain amount of drunkenness, but was, unlike many other commentators, not unduly censorious. In addition, rather than complaining about the problem, she set about trying to offer alternatives. She started a choir, which proved a great success, but she had to learn to cope with the fact that she never quite managed to assert her own authority over the men. They worked with great enthusiasm when the music that had been selected was to their taste, but if it didn't suit they simply walked out and went home – or back to the pub.

As she came to know the miners better, she began to discover that they were a genuine community, with their own traditions and their own organisations for mutual support. The savings club system was firmly established. For example, most families contributed 1s a week to the boot club – and then took it in turns to draw funds to buy new boots. There were occasions when all came together, some sombre, others joyful. Funerals were always great events for which the whole village turned out, forming a long procession, led by a powerful singing choir. Afterwards funeral cakes were handed out to every family, and to miss anyone out was considered a social disaster. Christmas was always a real community event, too. The celebrations began with a procession where King Coal, a huge block of coal garlanded with evergreens, was carried all round the village. What impressed the authoress more than anything was the sense of communal responsibility; the willingness of all to rush to the aid of any neighbour in trouble. She was less impressed by the houses, which were occasionally 'tidied up'– this appeared to consist of little more than plastering them with mud as a sort of primitive stucco. An attempt by 'The Master' to make improvements by suggesting paving over the village street and covering the open sewers was met with indifference.

Other commentators were less charitable, such as an 'Incumbent of the Diocese of Durham' who had scarcely a good word for the mining community, describing his village as 'a dark, dreary, dirty, smoky hole'. He did, however, remark that the houses were 'well ventilated', which it turned out simply meant that the front door opened directly into the living room/kitchen, from which the air had an unimpeded passage, before blowing straight out of the back door.

The Durham miners had their own customs and their own Christmas celebrations, when miners crowded into Sunderland or Durham to perform their sword dances. The men all wore white shirts decked out with coloured ribbons and were commanded by the captain, dressed in a cocked hat and a usually rather dowdy uniform. The collecting box was taken round by a clown or 'Betsy' who wore a fur hat with a dangling fox's tail. The similarity to the costumes worn by latter-day Morris dancers is obvious. Music was an important part of the miners' social life. The first mention of a colliery band appears as early as 1809, at which time it would not have been the familiar brass band we know today, but more a wind and woodwind ensemble. It was the introduction of valves in instruments such as trumpets and cornets in the 1830s that made it far easier to play tunes on brass and so began the period of change. Many colliery owners encouraged the formation of bands – and they became hugely popular. A tradition of band competitions developed. The first was held at Belle Vue, Manchester, in 1853, and attracted an audience of 16,000. It was not long before miners were organising their own events. The Northumberland Miners' Picnic was first held at Blyth in 1867 and bands from all the surrounding collieries marched on the town and then competed for prizes, one for local bands and another for bands from outside Northumberland. The example was followed in 1871 by neighbouring Durham with the first of what would become the famous Durham Miners' Gala. This was to become one of the major events not just for the mining community but also for a wide range of left-wing organisations. The bands marched with huge, colourful banners, often carrying socialist messages in later years, and the political speeches became as important as the music. The tradition has survived, even though the last deep pit in Durham has now closed.

South Wales was to become known for a different form of music-making. Mining began in the Rhondda valley in 1865, producing coal of exceptionally

The first practical mechanical coal cutter was introduced by Thomas Harrison in 1863: a circular cutting disc driven by compressed air.

Although there are no longer any deep mines left in the Durham coalfield, the 2009 Durham Miners' Gala proved to be one of the biggest ever, attracting a crowd of over 10,000.

high quality that became known as steam coal. Within fifty years the population had risen from less than 1,000 to 150,000, and fifty-six pits were opened in the valley. Men outnumbered women and so a number of all-male societies grew up, including the male voice choirs. They were to become famous for their musicality and travelled the world. A choir from the Rhondda took first prize at the Chicago World Trade Fair Eisteddfod in 1893 and the Treorchy choir was invited to sing for Queen Victoria at Windsor.

Not all of the miners' favourite pastimes were quite this innocent. Gambling on simple games like quoits and pitch and toss was very popular, but there were far more alarming activities than that. The 1842 Report on Children in Mines described a particularly vicious sport that was popular in Lancashire:

> The colliers are great fighters and wrestlers. On Christmas day I saw twelve pitch battles with colliers.
> Were they stand up fights? – No; it is all up and down fighting here. They fight quite naked excepting their clogs. When one has the other down on the ground, he first endeavours to choke him by squeezing his throat, then he kicks him in the head with his clogs. Sometimes they are very severely injured; that man you saw today with a piece out of his shoulder is a great fighter.

It is difficult to get an accurate picture of just what life was like in a colliery village in the nineteenth century, simply because most of those who wrote about the subject did so from a particular perspective and the colliers themselves had neither the leisure time nor the inclination to write their own stories. It is possible to get some idea of the living conditions from the housing that has survived from the period, but even that evidence has to be tempered with an understanding that the worst housing has long since fallen down or been demolished. It is, however, also worth noting that some owners at least provided really very decent housing for their workforce. Earl Fitzwilliam owned collieries at Rawmarsh and Elsecar near Rotherham, where the mines were remarkably free of gas and, as one visitor noted, 'so clean and commodious are the broad ways, in many parts, at least, that the ladies of Wentworth House sometimes go down to witness the operations'. Some of the houses the earl built have survived. Reform Row, Elsecar, consists of the standard pattern of two-storey houses, but built of good quality stone and to a very high standard. There are small walled areas in the front and gardens or yards at the back, accessed by an alley that runs behind the row of houses.

One feature that was lacking from virtually all miners' houses in the nineteenth and well into the twentieth century was a bathroom. This made life particularly hard for the mining community. There was no legislation insisting on pithead baths throughout the nineteenth century, so the men had to walk home, blackened with coal dust. The nearest they ever got to having a proper bath was a tin tub on the floor.

Housing in some other mining districts was very different from that of the coalfields. In Cornwall, in the late nineteenth century, many miners were building their own houses. A witness to the Parliamentary Committee on the conditions of workers in mines in 1864 gave a very full account of the housing conditions in the Camborne area:

> I think there are 800 houses in this town, the property almost entirely of the labouring miners, which they have built themselves and which they have on the lease of three lives.
>
> I am told that the aggregate amount of house property belonging to labouring miners in Camborne exceeds £50,000.
>
> 'Do you consider this an evidence of their prudent habits?' inquired one of the Commissioners.
>
> Of their prudent habits – yes – and of their temperate, orderly, and good conduct.
>
> Do you think the miners are more in the habit of living in their own houses now than there were thirty years ago?
>
> Yes, beyond all comparison.[2]

The cost of building was estimated as anything from £50 to £80. Costs were even lower in the country areas. Here, the miners enclosed patches of wasteland, building walls out of the boulders cleared in the process. Often they blew up the rocks with gunpowder, and as a result they got as much as 2,000 or 3,000 tons of good, hard granite per acre, which was available for building. The results can still be seen. I visited a pair of semi-derelict cottages near Botallack mine in the 1970s, which must have been typical of many others. The construction was quite crude, the walls built up of irregular, large granite blocks. One of the pair had a central doorway leading into a single, stone-paved living area. The large fireplace had a straight flue, leading directly to a short chimney. The first floor was open to the rafters and the slate roof. The second cottage had much the same layout, but was considerably smaller. R. Quiller Couch described a hamlet composed of similar cottages that he visited in the mid-nineteenth century:

> The population is mixed, but the miners invariably occupy the most exposed and worst built cottages ... surrounded by cesspools, broken roads and pools of undrained rain. The village of Amal-Voer is like a cluster of cottages huddled together on the top of a hill with scarcely space between them for access. The bedrooms are rarely more than one in each house, and open to the ceiling. This gives that appearance of space; but if the roof is slate, it produces great heat during a summer day.[3]

The landowners did quite well out of the system. They had the land cleared for them at no cost to themselves; cottages were built, gardens cultivated and after three generations they got the property back again. The houses in the towns were generally more substantial, but few had such amenities as running water. The women, however, had a solution for coping with their laundry. They took their washing to the mine and used the hot water from the steam engine. In areas where stone was less plentiful, cottages were built of cob and thatch, which was very cheap, though one old gentleman complained that his house had been very expensive to build. He reckoned it had cost him 50s.

One of the problems faced by the miners who built their own homes was that they had to look for land that was of no use for any other purpose, which often meant that it was a long way from the mines where they worked. Jory Henwood gave evidence to the 1842 committee:

I have known instances where men who had to remain in an atmosphere of 96°F, whilst at their employ, at a late hour of the night had to walk three miles to their houses. Some of these were too poor to be well clad; and after so frightful a transition of temperature and so long a walk against a fierce and biting wind often reached home without a fire and had to creep to bed with no more nourishing food and drink than barley-bread or potatoes with cold water.

Many commentators noted that drunkenness was far less prevalent in Cornwall than in other mining areas. The first temperance society was founded as early as 1805 in Redruth, but it was the spread of Methodism that had the greatest effect. Unlike the established clergy, the preachers were often working miners themselves who spoke to their friends, neighbours and workmates in plain language they could easily understand. Many of the preachers brought their message directly to the men at the mine itself. It was not unknown for the well-known miner preachers, such as Billy Bray and Dick Hampton, to address hundreds of miners. Inevitably there were hecklers, but the preachers generally had an answer. When a drunk turned up at one of Dick Hampton's gatherings at Godolphin mine and tried to persuade him to take a sip from his brandy flask, Hampton had just the right quotation at the ready: 'The drunkard shall come to poverty, and drowsiness shall clothe a man with rags.' Few of the preachers had received more than the most perfunctory education. Hampton recalled that he was sent at the age of 8 to learn reading from an old man who charged threepence halfpenny a week. The family had only enough money to pay for seven months, but he learned to read a psalter. Afterwards Hampton persevered on his own until he could read and understand the whole Bible. It was the patent honesty of these men, who received no pay for their preaching, that helped to convert so many to religion and temperance. It is surprising to find how, even in the most remote mining regions and against all the odds, literacy thrived. Few locations could be more remote than Wanlockhead in Scotland's Lowther Hills, yet there was a miners' library here as early as the middle of the eighteenth century.

The miners of Cornwall were able to find the money to build their houses simply because the 1850s and 1860s were boom times, with Britain leading the world in production of tin and, more especially, copper. Then disaster struck. Huge copper deposits were discovered by Lake Superior in America and the very rich Rio Tinto mining complex was opened in Spain. As a result, the world price of copper went into a sharp decline and the Cornish found it more and more difficult to compete. By the 1870s, productive areas such as Gwennap, which had once employed 10,000 men, saw mine after mine close until not a single one was at work. There was one redeeming feature. Many of the copper mines were already proving unproductive, but they had been sunk deep enough to reach new tin deposits below the veins of copper ore. The mines that had kept back some of their profits in the good years were now well placed to invest in developing the ores. Many, however, had simply distributed all their income as profits to the adventurers and now lacked the capital needed to develop new markets. Then things got even worse. The price of tin began to fall too, from £100 a ton in 1890 to just £64 a ton only four years later. Inevitably, more mines closed and the experts blamed it on the miners not working hard enough. Whether or not it was the miners' fault, they were the ones who suffered most as more and more faced long periods of unemployment.

The Cornish did not sit back and wait for things to improve. The industrial world was clamouring for more raw materials, especially metal ores, and trading was now a worldwide concern; no longer limited to supplying local needs. Mines of different kinds were opening all over the world and the West Country men found they had marketable skills to sell. The result was a mass exodus of miners from Cornwall. It was estimated that between 1871 and 1881 the Cornish mining population was reduced by a third. The migrating workers became known as the 'Cousin Jacks' and it was said that if you went down any mine anywhere in the world you would find a Cornishman at the bottom of it.

It was not the first time that Cornish miners had set off to work overseas. The grandly named Anglo-Mexican Mining Association was reported to exploit gold and silver mines, and offered what must have seemed small fortunes to those prepared to make the long journey. Working men were said to get £15 a month and mine captains £1,000 a year, compared with their Cornish rates of 50s and £80 respectively. The mines, it turned out, were poor, the management hopelessly inadequate and the conditions atrocious. Out of one party of forty-four miners, only eighteen survived. The promised high wages never materialised. Others did at least get adequate pay as they made their way across the Atlantic to work in the mines of Cuba and Brazil. Men were willing to leave their families behind for the promise of good pay. John Chynoweth of St Agnes was hired for six years to work in South America at a wage of £9 a month, but before he left he had a legal document drawn up appointing trustees who would receive £4.50 a month out of which they were to pay £2 a month to his wife and children, and they were also authorised to spend anything necessary on medical fees. The rest of the money was to be put in a savings bank.[4]

There were two major exoduses. The first was in 1849 when news of the gold strikes in California reached Cornwall; the second came when the Kimberley mines were opened in South Africa. Some returned home after an absence of many years. The Cornish went to San Francisco when it was little more than a few cabins, and to Johannesburg when it was made up of tents and tin huts. Near the centre of the latter was an intersection of dirt roads, known as 'Cousin Jack's Corner'. From 1890 to 1900 there were special trains to Southampton to take the migrating miners as they sought work around the world. In one family all nine sons went abroad and they were not untypical: one died in New Zealand, another in America, a third in Australia, and a fourth died from lung disease from working in South Africa.

The Cornish miners were not the only ones to suffer from a collapsing trade. In Derbyshire, for example, the lead mines also went into decline. In 1841 the industry employed 1,461; by the end of the century it had fallen to just 285. The Derbyshire men, however, were not in the same situation as the Cornish. Other industries, such as engineering, were developing in the area that offered alternative employment. The decline of the metal mines was to continue into the twentieth century, while coal mining continued, for a time at least, to thrive.

CHAPTER ELEVEN

UNION

The failure of the coal miners' strikes in the 1870s led to the steady decline and eventual disappearance of the Amalgamated Association of Miners as an active force. The grievances that had caused the strikes in the first place had not gone away, so it seemed that the only answer was to try a radically different approach under a new type of leadership. Unions and union leaders both appeared discredited, so in coalfield after coalfield the men entered into sliding-scale agreements. Wages were now to be tied to coal prices and all negotiations left to the Sliding Scale Committee.

Not everyone was convinced that this arrangement would work and some were totally opposed to it. It was all very well if prices stayed high, but without any sort of minimum wage agreement, a drop in prices could bring disaster, as union leader and politician Lloyd James explained with a good deal of vehemence: 'The present agreements they are going into on fluctuating market prices is a practical placing of their fate in the hands of others. It is throwing the bread of their children into a scramble of competition where everything is decided by the blind and selfish struggle of their employers.' Collieries were free to compete by cutting their prices simply because every time they did so, they were able to reduce the wage bill at the same time. The men who served on the committee were in the invidious position of trying to satisfy the demands of both the miners and their masters. It might all have worked if the system had been operating during an economic boom, but that was very far from being the case. The economy slumped, prices dropped and wages fell with them. Sliding scales that had been agreed were abandoned and new, less favourable ones introduced. With every adjustment of the scales, wages sank ever lower.

The new system was proving no more effective than the old in preserving a decent standard of living for the men and their families. It was thus not long before the old patterns were being repeated. Miners went on strike and looked for support among men in other districts. In 1883, Lancashire mine leaders promised to support the strikers of Staffordshire by raising a levy to provide them with funds. Handbills proclaimed the need for unity:

Shall we not buckle on our armour afresh and fight with greater vigour? The battle at present is being fought in Staffordshire, but if we are beaten there, we shall either have to fight in Lancashire or surrender. The Levy may be considered by some men very heavy, but the reduction will be threefold more. Who amongst our ranks would rather give the employers 2s than the men 8d?[1]

The rhetoric was fine but the results were disappointing. Many men were unconvinced by the argument that they should reach into their pockets today to help others, even if by doing so they might save themselves from worse conditions at some time in the future. The Staffordshire men never did get enough in the way of strike funds to hold out for long, and so the old, familiar story was repeated. They went back to work on the owners' terms at lower wages. With hindsight, it is easy to see that with no real sense of unity it was relatively simple for the employers to impose lower wages, area by area, just as the Lancashire handbill predicted. Hindsight, however, does not take full account of the conditions of the time. Communication was difficult and each region had its own loyalties and traditions, even, in the case of Wales, a different language from that spoken in the rest of the country.

The National Union was, in reality, very far from being national, drawing almost all its support from the north-east of England. It was moderate in outlook, committed to working with sliding scales and to co-operation with the owners. There was also a belief that it could have an influence on legislation and they claimed some of the credit for the passing of the 1872 act that had produced far better conditions for women and children, and established the rights to appoint checkweighmen. But there was nothing the unions could do to prevent the sliding scale moving inexorably downwards; they were no more able to fight the trend than the local unions had been. By the end of the 1880s there was only one word that could be used to describe the organisation of unions in Britain's coalfields: chaotic. Unions were losing members and in some regions they simply ceased to exist. There was no shortage of theories over what should be done to change things. Some favoured political action, joining the increasingly vocal groups advocating the new doctrines of socialism and communism.

One of these advocates had begun his working life as an 11-year old trapper boy at a mine in Lanarkshire in 1867 and had remained a collier for eleven years, eventually ending up as a hewer. In spite of having virtually no formal education, he taught himself to read very widely and even learned Pitman's shorthand. Armed with his new skills, he left the mine to work as a trade union activist. His name was James Keir Hardie. He led the miners on a strike against a lowering of wages, despite the advice of the official union, and it was in fact the very first strike ever held in that part of Scotland. Known as 'the tattie strike' because the men managed to get some sort of living by potato picking, it suffered the same fate as others of that time. But Keir Hardie's leadership was recognised and he was invited to become secretary of the Ayrshire Miners' Union. He founded a new journal, *The Miner*, and used its pages to campaign for a union for the whole of Scotland. When he began his union career, Keir Hardie was a conventional Liberal, but the events of those years forced a change of attitude. The crux came with a strike in Blantyre, where police were used to physically assault the strikers in a series of bloody confrontations. He began to have talks with the new advocates of Left-wing policies that included a meeting with Engels. He stayed clear of the Marxist camp but did go on to become one of the founding members of the Labour Party. It marked the start of what was to prove a long period of co-operation between the political party and the miners' organisations, though many years were to pass before it produced any meaningful results in terms of legislation.

James Keir Hardie (1856–1915) was influential as a leader of the miners and as a politician. In 1899 he became the first chairman of the Labour Representation Committee, which changed its name in 1906 to the Labour Party.

The Marxists and socialists were not the only ones who tried to analyse the economic forces to find out what had gone so badly wrong, and to ask what might be done to correct it. One answer that some miners' leaders came up with was that too much coal was being produced, and, therefore, the price was low. If less coal was produced then surely the price would rise again, and there was a simple answer to the question of how to reduce production: let the men work shorter hours. It was, as some saw it, a win–win situation: the owners got higher prices and the men got a decent wage for less work. The next step was to restart an old campaign that was dear to the heart of many miners – the eight-hour day. They even wrote a ditty that was sung by Yorkshire miners:

> Eight hours' work
> Eight hours' play
> Eight hours' sleep
> And eight bob a day.

'Bob' is an old name for shilling. The mine owners were unimpressed by the argument that had seemed so logical to the men.

In the event, the only way forward seemed to be a return to unions, but this time with a real attempt to work towards meaningful co-operation. A new leader emerged. Like Keir Hardie, Ben Pickard had started his working life as a young boy down the mine, but had soon become active in the union. He joined the Kippax lodge of the West Yorkshire Miners' Association and in

1858, at the age of 16, he was lodge secretary. He went on to become general secretary of the association and a keen advocate of co-operation between different regions. He pushed for amalgamation with the South Yorkshire Union, and by 1881 the two had been united to form the Yorkshire Miners' Association with Pickard as its general secretary. He was convinced by the conditions that he found when he travelled round the mining districts in 1887 that the only way forward was through more amalgamations:

> In most of the mining districts, low wages and starvation both in regard to food and clothing ruled. As I proceeded from one mining village to another and saw the destitution and impoverished condition of people, with the children going about barefoot and bare-legged up and down the little streets, I came to the conclusion that better things should be the lot of the mining population. The more villages I entered and the more information I obtained, confirmed me in the determination to rouse the people, not merely in Yorkshire, but throughout the country.

The following year, the price of coal finally began to rise again, and the Yorkshire miners put in a request for a rise in pay, which was rejected. Pickard was not about to repeat the mistakes of the past by calling for a local strike. On 10 September 1888, he sent out an invitation to the representatives of all miners' organisations that were not tied to sliding-scale agreements to meet to discuss ways in which they could push for a 10 per cent rise. Two weeks later delegates from Yorkshire, Lancashire, Derbyshire, Nottinghamshire, Leicestershire and North Wales met at the Co-operative Hall at Ardwick near Manchester. They passed a resolution agreeing to a demand for the 10 per cent rise, backed with a threat of strike action by every area if the demand was rejected. The timing was right. The slump was over and most of the regions agreed to pay up after a minimum amount of arguing. Only the mine owners of Derbyshire and Yorkshire held out. This time, divide and rule did not work. All the other areas at once began collecting a levy to support strike action and within days the raise had been granted. It could hardly have been a better start for the loose-knit association. As Pickard himself put it, 'Remember, by Unity we got the advance, and by Unity we may preserve it'.

The principal barrier to genuine unity was the existence of the National Union which, even if it was a spent force, still laid claim to the right to speak on behalf of the whole industry. Its weakness was highlighted at a conference in Birmingham in 1889, where the official's very moderate and non-confrontational approach was exposed to a challenge by the tiny Ayrshire Union's leader Keir Hardie, demanding support for the eight-hour day movement. In his speech to the conference he made his own views plain: that the militants had as much of a right to be heard as the moderates. It was clearly time for the movement to press forward to genuine unification.

Pickard called another conference in Newport on 26 November 1889. The location was specially chosen to reflect the new feeling of militancy, as it celebrated the fiftieth anniversary of the Newport Chartist uprising. They met this time with a series of successes to their credit, and, by handing out £11,000 in strike funds to obtain the wage rise for all districts, had proved the case for nationwide action. There was another good reason for choosing South Wales as the venue. It was here that the old association had been broken, where the owners were united in strong combinations and where unionism was at its lowest ebb, and sliding-scale agreements were strongest. They were definitely cocking a snook at the Steam Colliers' and Ironmasters' Associations. On the first day of the conference a formal resolution was passed to acknowledge what had already been accomplished in practice. The Miners' Federation of Great Britain (MFGB) was formed, which was to be the voice of most British miners for the next half century.

The MFGB at once took up the cause of the eight-hour day, which did nothing to improve relations with the north-east and the National Union. While this political objective was being pursued, the federation continued to push ahead for wage increases. As long as coal prices were high they had a certain amount of success, but as soon as prices began to fall again they had to abandon the political campaign to mount a defence against wage reductions, which were being proposed for virtually every coalfield in the country. At first, the owners concentrated their efforts on areas where the MFGB had not yet got a firm hold, using the tactics that had served them well in the past: picking off the weakest regions one by one. But it was obvious that if the aim of getting a general reduction in wages was going to be met, at some time they would have to face up to the federation. In 1893 the real battle began.

The MFGB was still a very young organisation, but it proved a lusty infant. The owners hoped to exploit any rifts that might occur, but the miners were now determined not to agree to a penny off the wages at any pit. The leaders must have been anxious: they had all been in this sort of situation before, fighting battles that they seldom won. But they put the question to their members: 'Will you agree to 25 per cent reduction in wages, or any part thereof?' The result was an overwhelming majority for fighting the cuts, 221 members voting 'yes' and 143,695 voting 'no'. The federation gave their answer to the owners, who responded with a lock-out. The struggle that followed was as bitter and bloody as any contest in the collieries had ever been.

Six weeks into the lock-out, word reached the miners of Featherstone, Yorkshire that poor-quality coal which had been stacked up at Lord Masham's Ackton Hall Colliery was being sold off. Trucks were being loaded by surface workers and 200 men marched to the colliery to confront the men they regarded as strike breakers. There was an argument, a few trucks were overturned and then the men left. The mine manager at once rushed off to Pontefract to demand police protection for the workers. Unfortunately for him, he arrived while the police were rather busily, and probably more entertainingly, occupied on crowd-control duties at the Doncaster races. The manager, however, soon found a very influential supporter in the person of Lord St Oswald, who was also a local mine owner and a Justice of the Peace. He also appealed directly to the chief constable, who suggested that as his force was not available they should call in the military. The suggestion appealed to Lord St Oswald, who doffed his mine-owner's hat and put on his magisterial robes, and issued the appropriate orders.

The soldiers eventually arrived at Ackton Hall to disperse a crowd that had already long since gone home. Nevertheless, they were deployed all around the colliery and inevitably men soon began to appear on the scene to find out what was happening. Having failed to find an angry crowd to deal with, the authorities had now contrived to create one. Tension mounted: a few stones were thrown at the engine house and a pile of timber was set on fire. The captain in charge was fortunately a sensible man who could see that his soldiers were making the situation worse rather than improving it. He went to talk to the unofficial leaders in the crowd, and it was agreed that if the soldiers left everyone else would also leave the colliery. It was a sensible and amicable agreement, and that should have been the end of the matter.

Captain Barker and his men returned to the local railway station, where they met a much less sensible gentleman – a local magistrate, Mr Hartley of Pontefract. He at once ordered the troops to return to the colliery. The miners felt betrayed and soon a much bigger crowd gathered, one that was in a far worse temper. Hartley at once read the Riot Act and ordered the crowd to disperse. They were in no mood to give in after recent events, and the order was given to open fire. The soldiers fired directly into the crowd and two men were killed and sixteen injured. But still people refused to disperse. Reinforcements were called in and eventually the crowd broke up. If the extra troops had not arrived, the situation could have been far worse and erupted into a full-scale battle, but, as it was, the whole affair had been disastrously and fatally mismanaged. The authorities had contrived to turn a minor incident into a bloodbath. It seemed as if the ideal of negotiated settlements was as remote as ever, and that brute force was still considered preferable to argument and reason.

The lock-out continued and it seemed that the other weapon that had always been successfully employed by the owners in the past would work again; families were becoming desperate as funds were exhausted. However, this time the story was to have a very different ending. Events like the shooting at Ackton Hall Colliery and stories of the suffering of women of children began to have an effect on public opinion. It was not as if these were men pushing to get higher pay, they were simply trying to prevent a reduction in what were already low wages, in what the politicians were telling everyone was now the richest and most powerful nation on earth. Sympathy with the miners was converted into practical help: money, food and clothes began to pour in for distribution by the federation. Soup kitchens were set up to make sure that everyone had enough to eat. Now, instead of a trickle back to work by the miners, it was individual collieries that began to give up the fight and took men back on reasonable terms. Every colliery that reopened was a double victory for the men: the families no longer had to receive strike pay and they could contribute towards helping others who were still locked out. Rather than getting weaker, the strikers' position was actually improving.

In the end, it was government intervention that brought this contest to an end. At first it was the Board of Trade that tried to arbitrate between the two sides, but when that failed it was the prime minister himself, William Gladstone, who stepped in. He appointed an official arbitrator, not a civil servant but the Foreign Secretary, Lord Rosebery. He was not a man either side could ignore, and they agreed to abide by his decision. On 17 November, sixteen weeks after the start of the dispute, he delivered his verdict. The men were to return to work under the old rates – there would be no reduction in wages. Thomas Ashton, the MFGB secretary, described how the miners celebrated what was for them a complete victory with 'singing, dancing, shouting, laughing and crying for joy, and in several districts the church bells were set ringing to celebrate the great event'. It had been a taxing time for Lord Rosebery but he felt that it was a job well done. He wrote in his diary: 'Dined alone, very tired. But it would have been a good day to die on.'

The value of the federation had been demonstrated, but it did not yet represent all miners. One vitally important area, South Wales, was still firmly tied to sliding-scale agreements. But they were discovering, as others had done, that it was easier to slide down a scale than climb up it. They were suffering a steady erosion of conditions and pay, and their patience finally gave out. They demanded a 10 per cent rise and an end to the old agreement. The owners' response was immediate. The men received their notices and a total lock-out began in April 1898. The circumstances were in many ways exactly the same as those that had faced other regions in 1893, but with one important difference. The Welsh miners were standing alone. The way the situation developed also followed the pattern of the earlier lock-out. The military were called in to break up any meetings of the men, even though the peace had neither been broken nor threatened. Both sides settled down to a war of attrition. It was just a question of whose nerve broke first or whose funds ran out. It was now that the lack of a wider organisation and substantial funds was felt, and the result was inevitable. The men had to go back on the old terms and the contrast between the two lock-outs could scarcely have been starker. Nevertheless, the lesson had been learned. The following year the Welsh applied to join the federation. Edward Conway, seconding the motion to admit the Welsh, spoke for the executive as a whole and perhaps especially for Ben Pickard, who had fought and argued so long for a national union:

We have made many attempts and many tries to get our Welsh brethren to join.
I must admit that we had almost given up hope. But now the time has arrived;
and I am exceedingly pleased, as one of the founders of this Federation, that this
time has arrived and our friends have now joined us. It will help to make us one
of the most powerful federations in this country that has ever been seen.

There was no better argument for those who favoured a national federation
than to point to the successes it had already achieved. It had already been
growing steadily, from a modest beginning: 36,000 members in 1888, 96,000
in 1889 and by 1893 it had passed the 200,000 mark. Now, with the addition
of the Welsh miners, it was stronger than ever. The north-east clung to its
old National Union for a time, even though it was patently neither national
nor especially united. First Northumberland and then Durham joined the
federation, and by 1908 the days of two competing unions were finally ended.
The miners spoke with one voice and that year the voice was heard loud
and clear, not for once in protest but in cheering for a measure that many
thought they would never see as part of the law of the land. The government
finally passed the Mines Eight Hours Act. It had been a long battle to get the
legislation through; promises of action were made but nothing ever seemed
to happen. The National Union had led the fight at first and then the MFGB
had taken up the cause. Now it was a reality. It was not everything the miners
had hoped for. They had argued that the eight hours should start from the
time they set off down the pit to the time they emerged at the end of the
day. Instead, the hours were to be timed from arrival at the coalface to start
work to leaving again. In the big mines there were often long distances to be
travelled from the bottom of the shaft to the face and back again, and there
could be long waits for the cages back to the surface. All of this meant that the
actual time spent underground could be well over the eight hours. However,
that was far less important than the fact that the grand prize, which had been
fought for for so long and so hard, had finally been won.

Inevitably, the bill had been opposed every step of the way by the mine
owners, and the fight was not abandoned even after the bill had been passed.
Lord Newton raised the issue in the House of Lords on 5 August 1909
when he asked whether the government was going to amend the act: 'In
putting this Question I merely desire to observe that this Act only came into
operation on July 1 last, that the reduction in output on the average amounts,
I understand, to something like nine per cent.' He received short shrift from
the Lord Chancellor: 'I am sorry to spoil the picture which was drawn by the
noble Lord, and which I think was more entertaining than accurate.' Over
the next few years, there were numerous questions raised suggesting that the
industry was suffering and there were constant demands for statistics, which
the government always answered by replying that they did not actually have
any reliable figures, because no one was collecting them.

The miners hardly had time to take down the bunting from the parties celebrating the Eight Hours Act before the next dispute began. It was to turn out to be a bigger strike than any that had ever hit the country in the past. The federation called for a show of strength and unity, and the result was that almost every coal mine in Britain stopped work. The issue was the call for a minimum wage. This was of huge importance to the vast majority of the workforce who were paid by the amount and value of the coal they produced. Two men could go down a pit, work with equal energy and expertise for exactly the same amount of time, and yet end the day with very different wage packets. The difference had very little to do with the men themselves, and everything to do with the geology of mining. One man would be sent to a stall, which was high and wide, which gave access to a seam of first-rate coal. At the end of a week he would be delighted by his earnings. The other might go to a place where the seam was no more than 2ft high, where the coal he won came in unprofitably small pieces, mixed with rock and other rubble. The effort of working in such a place was immense and he faced far more difficulties than the miner in the wide seam. Such places were often dangerously insecure and time had to be taken out to insert props to prevent the whole seam collapsing. At the end of the week the second man's pay packet would scarcely be big enough to keep his family from poverty and starvation. There were also grievances about the ways in which work was allocated. It was all too easy for supervisors to settle old scores and managers to punish troublemakers by sending them to what were known as 'abnormal places'. But the greatest resentment was always the unfairness of two hard-working men in the same occupation receiving vastly different wages. Motions were sent to the coal owners and they were turned down.

The strike began in February 1912. The government was alarmed. Coal quite literally fuelled the country's economy. As well as the old users of coal, a new source of power was becoming increasingly important: electricity. Britain's first power station was opened at Deptford in 1889 and the electricity was generated by four huge 10,000hp steam engines. It may have been a new source of power, but it still gobbled up great quantities of coal and, as time went on, electric power stations were to become the coal industry's best customers. The government was not about to see the whole country brought to a standstill while the two sides wrangled. On the evidence of past disputes, they were unlikely to reach agreement until one side had caved in. So Parliament passed its own minimum wage bills, which managed to satisfy neither side. It drew back from the idea of a national minimum wage, but instead called for deals to be organised at a local level, through boards that would be made up of representatives of miners and employers under an independent chairman. The owners were not happy with anyone meddling in their affairs, and the MFGB saw it as a threat to their newly discovered sense of national unity. Parliament was not in the mood to listen to complaints from either side, however, and the bill duly became law.

The federation found itself in something of a dilemma. Accepting the act meant going back to the old system of regional deals with bargaining over wages being handed over to the new and, as yet untried, boards. On the other hand, opposing the act would set them in opposition not only to the owners but to the government as well. Whatever their misgivings, there was a strong feeling among many unionists that this was a fight they could not win, and they risked losing the very real gain of having an established minimum wage in place. The federation reached a far from unanimous decision that they would co-operate and try to make the new system work. They had not given up all their rights: they were still able to negotiate on wages in general – only the minimum wage was taken out of their hands and given to the regions. There was no denying the fact it was a retrograde step as far as the long fight for national unity was concerned. However, there were very real gains to celebrate. The boards did their jobs and managed to set reasonable wages for all coal face workers, and that brought thousands of families out of abject poverty. It was a hastily cobbled together piece of legislation, designed to avert a short-term disaster, not to deal with a long-term problem. The surprise is that it worked as well as it did, even if it never really satisfied anyone.

The federation could look on a period of success on the national scale, but there were always going to be local issues that were less easy to settle. The most famous, or infamous, took place in South Wales in 1910. The various mine owners of the Rhondda had formed a group known as the Cambrian Combine, committed to acting in unison during disputes. One mine, the Naval Colliery Company, had opened a new seam and its development and the rates that would be paid depended on how efficiently it could be worked. When the results came in, the company declared that the men had deliberately been working slower than usual to get better terms. The men replied that it was a difficult seam to work because of a layer of rock that ran through it, and as they were paid by the amount of coal they produced not the hours they put in, they had no reason to falsify the position. On 1 September that year the company posted lock-out notices, not just barring the way to the seventy men working the new seam, but to all 950 miners.

The Ely pit now came out on strike, and the Combine called in strike breakers from other areas. At this point, the federation stepped in and balloted all 12,000 men working in the Combine's mines. An attempt at conciliation failed and every pit was closed with one exception, Llwynypia. Pickets arrived to try to prevent workers going down and the scene was set for violent confrontation. The union leaders tried to keep the peace but feelings were running high and some of the picketers began hurling stones at the pump house. The police were called in and, after a good deal of fighting, baton charges drove the men back into the square at Tonypandy. The next day there was a demonstration in Tonypandy that was violently broken up by the police. Now the story took a turn that was to earn it a special place in the history of the mining unions.

The chief constable of Glamorgan and the manager of the Cambrian Combine sent in a joint request for military reinforcements. It was passed on to the Home Secretary, Winston Churchill. He was reluctant to agree, but sent a detachment of Metropolitan Police to the area, and arranged for troops to be sent to Wales and put on standby in Cardiff. He informed the unions of his decision, but the local magistrate demanded that troops should be sent to the area. The Home Office agreed. A bad situation was made worse, and rioting broke out in Tonypandy, with many shop windows being smashed. The local police remained on duty guarding the mines, and it was only the Metropolitan Police who were available to deal with actual law breaking. By the time they arrived, several hours after the trouble started, the situation had deteriorated even further and the troops were called in. By the time calm had been restored, it was estimated that eighty police and 800 locals had been injured, and one miner had been killed.

To the miners of South Wales, Churchill was and remained the arch enemy; the man who had called in the troops from England to attack local miners. Nothing that he did in later life, not even in the years when he led the country in the Second World War, would ever quite reverse that opinion. The criticism is not altogether fair. He had been reluctant to use troops – it was the local magistrates and police authorities that had insisted. And if the local police had been used to help quell the riot instead of being left to guard the mine owners' property, which was not actually being attacked at the time, the whole matter could have been resolved without the military. However, whatever the truth of the matter, Tonypandy will always be remembered as the place to which Winston Churchill sent English soldiers to attack Welsh miners.

In spite of setbacks, the MFGB was proving to be a success. The miners took the search for unity a stage further, when they began talks with other unions to see if they could work together. The result was the Triple Alliance between the miners, the National Union of Railwaymen and the Transport Workers' Federation. There were obvious advantages to such a coalition, but the real test would be whether such a loose alliance could ever have the same coherence as one between men who did the same jobs and shared the same traditions. When it was eventually put to the test, the agreement was to prove all too fragile.

So far, we have been looking at miners' unions purely in terms of coal mining, simply because unionism never really took hold in the metal mines. That is not to say that there were no disputes, but there were far fewer, largely because of the system of regular settings, where the men bid for the work and pay rates were set. The steady decline of the copper mines in Cornwall, and the willingness of the men to leave and try their luck anywhere in the world where there were pits to be dug, also had its effect. Although the copper mines in Cornwall were suffering, across the Tamar, in Devon, one great mine complex was thriving, for a time at least: Devon Great Consols. In the middle of the nineteenth century, miners hit upon a rich lode, in places as much as 40ft thick. One commentator wrote that it was so magnificent that it shone like gold and in the early years it was to prove as valuable as gold to the investors. There turned out to be another valuable mineral in the same mine, arsenic.

The arsenical ore was brought to the surface, where it was broken into small pieces, before being roasted in a calcining oven with a tall stack, which it was hoped would safely deal with the noxious fumes. The final stage of preparation was to take the calcined ore and reheat it in a refining furnace, where the vapour was cooled and deposited as crystals on the tiled floors of brick chambers. From there it had to be removed by hand. This was lethal stuff: it was estimated that one-sixth of a teaspoon was enough to kill a man. The workers who had the unenviable task of removing the crystals wore rough masks over their mouths, earplugs, and sacks wrapped round their ankles and feet. By 1871 they were producing half the world's arsenic.

Devon Great Consols paid huge dividends for many years to the shareholders, but wretched wages to the workers. The average for the latter was £3 12s a month, but a month was reckoned at five weeks. There was one strike in the early years, in 1850, when 200 men came out and were promptly sacked. By the 1860s things were very much worse. Wages remained largely the same, but the price of provisions was rising at an alarming rate. For the first time, something like a union was formed in 1865: the Miners' Mutual Benefit Association. They had a long-standing grievance about the employment of unskilled workers in the mine, simply because they slowed up the work of the experienced men who were working the tribute system, where what they were paid depended on how much ore they could raise. They attempted to introduce a rule that the miners should have a part in deciding what everyone should be paid, at which point the owners responded by locking out all the men who were known to be in the association.[2] This involved most of the mines in the Tamar valley.

The situation deteriorated rapidly. On 26 February 1866, a group of locked-out men managed to grab one of their former colleagues who had gone to work. They set him astride a pole and carried him through the local mining town of Gunnislake, where he was greeted by crowds shouting 'blackleg'. The local mine manager was terrified and called in the police who arrested the men involved. The local magistrates sent the men to prison, which inflamed the situation and brought forth an angry response from around the region. Mine agents and captains started receiving threatening letters and more and more police were drafted; eventually there were 130 constables on duty to protect Devon Great Consols.

The following week saw a further deterioration in the atmosphere round the mine and the owners began to fear that the pumps would be stopped, which would have inundated the workings. Having used up the resources of the police force, the magistrates now called in the military and 150 troops arrived from Plymouth. Just in case that wasn't enough, 150 extra special constables were called in. Saturday 3 March was the day set aside for the bargains to be made for the next set of underground workings. An estimated 2,000 miners gathered outside the count house, watched by the 300 constables. The prices offered by the company were read out one by one and each was met with complete silence. Noise only erupted when the director, W.A. Thomas, told the miners 'to go to work like men'. They shouted him down.

It was a brave but ultimately doomed effort. With so many miners unemployed in other parts of Cornwall, the company did not have too much difficulty getting men to keep the mines open. A trickle back to work became a flood as men returned on the company's terms, and the association was effectively finished. As it was, the glory days of the famous mine were already over. After a long period of decline Devon Great Consols went into liquidation in 1901, and 351 men were put out of work. Over the years, the mine had sold over £3 million worth of copper ore and over £600,000 of refined arsenic, which would be a huge sum at today's prices. Now it was all over: in 1903 pumping stopped and the assets were all sold. Unionism never rose again in the mines of the south-west; the industry and its profits were sinking from year to year. Throughout the latter part of the nineteenth century and into the beginning of the next, men had fought hard for better pay and conditions. Soon, many of them were to become involved in a far grimmer and bloodier conflict.

CHAPTER TWELVE

TWO WARS

A t the outbreak of the First World War coal production in Britain was at its height. Figures for 1913 put total production at 287 million tons, of which a third was for export. War changed the position dramatically as coal exports virtually disappeared. There was, however, a new form of employment available. All over Britain men were caught up in the patriotic fervour and rushed to join in the glorious war that was scheduled to end, with triumph, by Christmas. The reality was very different and soon the war became an almost stagnant battle on the Western Front, as the armies dug in for miserable years of trench warfare, where advances and retreats were measured in yards rather than miles. Attacks across no-man's-land resulted in huge losses for minimal gains. But another war was also being fought: an underground war. It was a war that was largely fought by miners.

Recruiting officers went round the colliery districts to sign up men for the tunnelling companies, offering them 6s a day. Altogether some 40,000 miners were recruited to take on the work of tunnelling under the enemy lines to plant mines. Typically, the men worked a vertical shaft down to a depth of anything from 20–30ft, and then set off in a straight line towards the enemy trenches. It was a desperately slow business. There were usually four men working at the face in relays and two more pulling out the trolleys of spoil, shoring up the sides and looking after the air pumps. The men would have found the work difficult and uncomfortable, but no worse than what they were already used to in the mines. They did, however, face an added danger. While they were tunnelling towards the German lines, the Germans were also tunnelling towards theirs. They had to be constantly listening out for any noise of activity nearby and occasionally they had some unpleasant shocks. Lt Geoffrey Boothby was an officer in the Royal Engineers. Typically, unlike the NCOs attached to the tunnellers, he had no experience of engineering of any sort, and had just started a medical degree at Birmingham when he volunteered.

In 1915 he had just taken over a tunnel from the miners who had declared it to be quiet and peaceful. It was officially inspected, and the inspecting officer went, out of curiosity, to look at an older working where, to his immense surprise, he discovered a German telephone. It was obvious that the enemy had broken through at some time and Boothby, together with another officer and a corporal, went down to investigate where the breakthrough had occurred. The tunnel was typical: 'our galleries are just like a square tube in the earth, going down straight and then shooting out and running parallel to the surface of the ground about twenty feet underground. These galleries measure four feet by three, so you're pretty cramped. In this old gallery there was about eighteen inches of water.'[1] The Germans had pumped water out of their section into the British tunnel, so that the noise they made sloshing through would give them advance warning that they were coming their way. Eventually they found where the Germans had broken through and discovered the remains of a telegraph system. They decided to explore a little further, but rather cautiously as they only had one rifle between them. They reached an area where the tunnel turned at a right angle, at which point someone shot at them. The German was in a safe position: he didn't have to expose himself, simply poke his gun round the corner and fire at random into the tiny tunnel, with a fair chance of hitting someone. Boothby had only one option: 'I beat an absolutely panic-stricken retreat, dashing through the water, bent double to avoid the roof, and crashing from side to side. It was, you realise, quite dark.' Back in the safety of their own section, an underground battle got under way as they began preparing a charge to blow up the tunnel, working flat out and piling up sandbags to protect their own position: 'The sandbags had to be dragged through water to us, and consequently we were absolutely wringing, and then the heat and the bad air, phew! The object of this haste was to get our charge off before the Germans got off theirs and buried us.' They won the race. A year later Boothby went to investigate another suspected German breakthrough. This time his luck ran out: a mine was exploded, burying him. His body was never recovered.

When explosions did occur, there was always a very real problem from gas in the tunnels, and the rescue work was entrusted to special teams equipped with breathing apparatus. One of the leaders and trainers of these teams was Sergeant Clifford, who had begun his working life assisting his father in the North Staffordshire Mine Rescue Service. All these men were engaged in a war quite unlike the one fought above ground; here, the inherent dangers of mining were increased by the ever-present possibility of meeting the enemy in the dark. No one ever recorded just how many hand-to-hand battles were fought far underground, but the work of the miners was undoubtedly of immense importance. A well-planned and well-executed attack underground could cause far more devastation than any artillery barrage. The most dramatic example came during the battle for Messines Ridge in 1917.

Work on tunnelling began eighteen months before the actual offensive, which in itself is a measure of the stalemate that was the Western Front. It was estimated that 8,000m of tunnels were dug, only one of which was ever discovered by the Germans. A total of twenty-two charges were laid and when everything was ready the officer in charge, General Plumer, wryly remarked: 'Gentlemen, we may not make history tomorrow, but we shall certainly change the geography.' He was not mistaken. At 2.50 a.m. on 7 June the artillery began to bombard the ridge. This was normally the signal that there was to be an assault so the Germans rushed to their defensive positions, with no idea that they were in mortal danger, not from the enemy advancing towards them, but from the mines right beneath their feet. That was where they were, peering out towards the British trenches when the charges went off at 3.10 a.m. It was the greatest explosion of the war, and was even heard by Lloyd George in Downing Street. It blew off the entire top of the ridge, and in the process killed some 10,000 Germans.

The largest of the mines exploded by the British under the German lines at the start of the Somme offensive left a 430ft crater at Spanbroekmolen. It has now been landscaped to create a 40ft-deep lake, known as the Pool of Peace.

The miners who went to Belgium also played an important part in the war, as did those who stayed behind providing the essential raw material of industry: coal. Very early on in the war, the government held discussions with trade union leaders who agreed a truce on industrial disputes. It was assumed that pay disputes would be settled amicably, but faced with steeply rising prices and a refusal to negotiate, strikes were called in the engineering industry. Once again government and unions met, and this time the result was the signing of the Treasury Agreements, which effectively gave up union rights to negotiate and strike. Key industries were also covered by a new piece of legislation, the Munitions Acts, which went further by making strikes illegal, and also authorised 'dilution', the employment of unskilled workers in skilled jobs.

Relations in the mining industry had never been good, especially in South Wales. The mine owners soon lobbied Parliament with their ideas of helping the war effort, which included abandoning the Eight Hours Act, reducing holidays and increasing productivity. There was no mention of sacrificing profits. The miners did not sign up to the Treasury Agreements, thus early in 1915 the South Wales men put in a wage claim that was rejected and in July they came out on strike. Theoretically, all 200,000 men could have been jailed, but that was hardly feasible. In the event, the demands were met. But now the government took tighter control, removing all negotiating rights from the owners and passing responsibility to a 'Coal Controller'. Collieries in the rest of Britain were soon brought into the scheme.

Over the next couple of years resentment grew. In 1916 Lloyd George himself came down to Cardiff to try to pacify the Welsh miners. It was generally seen that while the cost of living was rising and profits were increasing, wages were remaining static. In South Wales the miners again asked for a wage increase, and the whole matter was talked over by the War Cabinet.[2] The committee, chaired by the prime minister, was given a report on the situation:

> The Coal Committee stated that the men were out of hand, and the leaders were not leading, but were being pushed … the men based their claim for the present increase on the fact that the cost of living had risen during the war by 83 per cent. They also alleged that 50 per cent of the miners had to pay for tools, the cost of which had gone up during the war.

There was agreement that no one wanted to concede the rise, not only because they did not consider it necessary, but also because it was felt that it would encourage militancy in other industries. On the other hand, they could not afford the loss of coal production for any length of time, and if a strike was called it would be unlikely that it could be settled quickly. The result of a lengthy strike would be 'disastrous to the conduct of the war'. It was agreed that 'they had no option but to authorise the Coal Controller to negotiate up to a maximum of 1s 8d per day and 10d. per day for boys'. The eventual settlement was for 1s 6d, and 9d for boys under 16.

One aspect of government policy was vigorously resisted by the men; dilution. This was understandable. Mines were (and are) dangerous places, and safety depended to a large extent on mutual trust based on experience and a sense of community. No one was keen on the idea of relying on outsiders. Every miner was aware of what could happen to any of them at any time. Disasters made the news, but what seldom got reported were the accidents that happened frequently to individuals. Recently I visited the graveyard at Trinity Chapel in the Rhondda. The only memorial to a major disaster was for those who were killed when a dam broke at Cwmcarn, an event that had nothing to do with mining. However, all around it were the headstones of men who had been killed down local mines, not in big explosions, but in a series of individual accidents. Statistics tell the same story.[3] In 1913, 1,753 miners died in accidents and 177,189 were injured. Of these, 462 were killed in explosions that also injured 131. That compares with 620 who died and 62,094 injured by what were described as 'falls of ground'. The year 1913 was particularly bad for explosions: the previous year the numbers had been 124 killed and 145 injured, while the accident rate overall was much the same. Nevertheless, it was the explosions that had the biggest impact. The other accidents were spread all round the colliery districts of Britain, while an explosion affected just one pit and one community. This was the other war that the miners fought in the early part of the twentieth century.

There was an ever-growing understanding of the factors involved in explosions. Firedamp or methane had been recognised as a major cause since the eighteenth century. Ways of preventing it were also understood. The first essential was good ventilation. The use of powerful fans capable of extracting 200,000 cubic feet of air a minute had made a great improvement, but was only really effective if the air was carefully channelled throughout the workings. At the start of its journey, the air coming down the shaft was pure enough for naked lights to be used with safety, but gas would often be collected as it moved down the passageways, so great care had to be taken to ensure that none of the return air ever got mixed with the intake air. It only needs 6 per cent of the air-gas mixture to be methane for it to be explosive. The other obvious measure was to keep naked lights away from mines. One notable improvement had been the introduction of battery-powered electric lights for miners, but by the beginning of 1913, of roughly 750,000 lamps in use, only 11,000 were electric. The only problem with the electric lamps was that they gave no indication of the presence of gas, but that was easily solved by having just one standard safety lamp for testing purposes. Electricity was also coming into use for underground haulage, with a resulting increase in safety.

Until late in the nineteenth century it was generally assumed that firedamp was the only cause of explosions, but experiments found an equally lethal culprit, one that was present in all kinds of mines: coal dust. The problem is largely caused by the steps taken to keep the mine free of gas. The ventilation system produces powerful air currents that lift the dust and keep it suspended. Also in its long passage through the mine, the air gets hotter, and consequently drier. The danger depends to some extent on the nature of the coal itself: anthracite dust, for example, seldom if ever explodes, while dust from steam coal is extremely volatile. Dust arises from the different processes, from blasting and drilling to filling tubs, and the increased use of mechanical cutters in the early 1900s exacerbated the problem. There were no agreed methods of controlling the dust in the early years, other than by using water sprays. But as with gas, if there is no flame or spark, then there would be no explosion. Sometimes sparks were unavoidable: a miner may hit a rock instead of coal with his pick, or falling stones crashing together can have the same effect, but the risk could be minimised. The twentieth century brought a number of government measures designed to increase mine safety, of which the most important were set out in the Coal Mines Act, 1911.

The act was remarkably thorough, dealing with the problem under different headings. It laid down that a mine should be properly ventilated and that nowhere should be considered fit for work if the air contained less than 19 per cent oxygen or more than 1.25 per cent carbon dioxide. Qualified inspectors were to test for gas at regular intervals. Safety lamps had to be kept in special lamp rooms and had to be inspected after every shift. Miners were to be checked before going underground to make sure they had no smoking materials or matches. Electricity had to be cut off when gas was detected and explosives had to be properly stored and inspected. The coal dust problem was rather less stringently dealt with, the act merely saying that it should be kept down 'either by way of water or otherwise'. The final section was the most important. It specified that no shift should be allowed to start work beyond the station at the entrance to the mine where the men were checked out before it had been declared safe by authorised 'firemen, examiners or deputies'. It was all very sensible and took account of the latest research into mine safety. Yet disasters still occurred. What went wrong? Looking again at those statistics for 1913, it turns out that the very high death rate for that year was mainly due to one appalling disaster.

The explosion occurred at Sengenhydd Colliery on 14 October 1913 and 439 men lost their lives. It was thought that a rock fall had probably released a pocket of gas, which had then been ignited by a spark. There was an official inquiry, though given the level of destruction it was never going to be possible to discover exactly what had happened. But several very disturbing facts did emerge. It was known to be a fiery pit. There had been an earlier explosion with loss of life, and in 1910 there was such an accumulation of gas that the whole mine had to be closed for four days. In mines with such major problems certain things could be done and should have been done, but the mine manager, who was also the agent for the owners, did none of them. The rules laid down under the 1911 act were ignored. Normally in a pit with such bad conditions, there should have been a system of reversing the air flow to make sure the whole of the underground area was clear of gas. There was not. There should have been a system in place for keeping down coal dust. No such measures were taken. Various regulations concerned with the correct way of appointing officials to inspect lamps, the need to carry out regular inspections and record the results were simply ignored. Electric signals that were used to control underground movements had bare wires that were likely to spark, even though this danger had been specifically pointed out to mine owners in a Home Office circular just a few months before.

A case was brought against the manager and the company by the divisional mines inspector for Wales on behalf of the Home Office. The manager was found guilty of six charges, three of which were technical rather than likely to have contributed towards the explosion. The others were either serious or very serious in the view of the inspector. The magistrates then handed out their sentences: the manager was fined a total of £24. The headline in the local Labour paper said it all: 'Miners' Lives at 1s 1¼d each!' The company was held not to be liable in any way at all, 'as they had properly appointed a manager, not interfered with him, and made all necessary financial and other provisions, and the company had no knowledge of the offences'. It is not difficult to imagine what the widows, orphans, friends and colleagues thought of the court and its findings.

Professor Jevons, who had originally studied as a geologist but went on to become a Professor of Political Science at the University College of South Wales, was scathing in his account of the lamentable proceedings. He pointed out that the system for reversing air flow should have been in place by 1 January 1913, but the manager had asked for and been granted a four-month extension. Then another extension was granted up to 30 September. That date had been reached and still nothing had been done; a fortnight later the explosion ripped through the mine. It was not the only example of negligence. As Jevons wrote: 'Why are these laws passed if Parliament does not mean them to be enforced?' Why indeed!

One major problem was identified. There were too few inspectors, even though following the 1911 act the numbers had grown from forty to eighty-two, of whom eight were specifically dealing only with quarries. It sounds quite a lot, but Britain in 1913 had 3,289 working collieries. It was impossible to do thorough investigations of every pit on a regular basis, particularly as the inspectors were also required to investigate accidents, which could be very time-consuming. In practice, the inspectors did random checks. They would go down a mine and look at one particular safety aspect, and if that was satisfactory they assumed everything else was as well. William Brace, President of the South Wales Miners' Federation, expressed the union view:

> They have to cover too wide an area. By the time an inspector leaves his residence and gets to the collieries a long way up the valley, by the time he has examined the plans and read the fireman's reports, a substantial part of the working day has gone, and there remains only three or four hours for his underground investigation. We want a real inspection. There is not a colliery in the country but ought to be inspected once a month, and by inspection I do not mean that a man should turn up for a few hours and run round the workings ... that we should have, in fact, a thorough inspection so that men will be able to feel, after it is all over, that their colliery is in a safe condition.[4]

Those words were written in October 1914 and the government by then had very different preoccupations. The war against Germany ended in 1918; the war against the dangers of mining still had many years to run.

CHAPTER THIRTEEN

A TIME OF STRUGGLE

It was hoped when peace came that better times would return, with growing demands being put on industry, which in turn could only mean a greater demand for coal. It was certainly hoped that the vital export market would flourish again, but there was a new barrier to overcome, one created by the victorious nations themselves. The Treaty of Versailles demanded punitive measures against the losers, calling for reparations that included Germany supplying cheap coal to other countries. No one seemed to have noticed that however pleased with themselves the victors might have been, they were actually inflicting serious damage on their own coal industries. However, for the time being the economy seemed to be booming. The coal industry was thriving and still under government control, so the federation felt that this was a good time to push for an even greater change. They wanted the industry to be nationalised and, for good measure, they wanted a 30 per cent pay rise and a six-hour day. It is doubtful if anyone seriously thought that all their demands would be met and when the inevitable rejection came, members were balloted and voted overwhelmingly for strike action. At this point, Prime Minister Lloyd George suggested a compromise. He would set up a commission to look at pay and conditions, investigate the idea of public ownership and, unlike during earlier inquiries, the miners would be fully represented. The owners and the federation agreed.

At first things went well, and when an interim report came out with a compromise position on wages it was accepted by both sides. On nationalisation, the commission was unsurprisingly split between the owners' and the miners' representatives, but the chairman Sir John Sankey came down in favour of the idea of public ownership. Lloyd George might have been prepared to accept the idea, but he was in a coalition with the Conservative Party and they were never going to agree to such a socialist notion. A compromise was put forward for consolidating the different regions, but that plan was rejected by both miners and owners, so in the event nothing at all was done. The federation joined forces with the Labour Party and the TUC to run a campaign with the title 'The Mines for the Nation'. The plan was to gain public support for nationalisation, but the issue that had seemed so important when the economy was doing well faded away as the country started sliding towards a major slump. There were then the more familiar problems to be dealt with in the form of rising prices and falling living standards.

The miners decided that it was time to reinvigorate the Triple Alliance with the railwaymen and the transport workers of 1920. So when the miners went on strike for better wages, they only needed the other two unions to threaten to join in to cause a rapid climb down by the government. They offered a temporary rise that would last for six months, during which time the whole matter would be considered again. This was agreed, but it was only a stopgap and by now the government had had enough of dealing with these problems themselves. They ended the system of wartime controls and prepared to hand the running of the mines back to the original owners. Relationships between the two sides had not been good before the war, and they were certainly no better now.

The situation in the mining industry was worsening and the owners came up with their usual remedy: the men must take a cut in wages, which they refused to do. When looking at the history of mine disputes over the years, there is an inevitable sense of déjà vu. The same scenario is run and rerun. Times are bad, and the only solution is to cut wages; this is resisted, a strike is called and one side or the other eventually gives way. Lord Birkenhead, who had been involved in many of the talks about the industry, noted sourly of the miners that 'I should call them the stupidest men in England if I had not previously had to deal with the owners'. The war of attrition began on 31 March, the day that the government gave up control, at which point the owners locked out the miners. The government promptly declared a state of emergency, and troops were moved into the coalfields and machine guns placed at pitheads. The situation was grim, but this time, the miners felt they had shifted the odds in their favour, thanks to the Triple Alliance. The railway and transport unions agreed to come out in sympathy, with a strike scheduled for Saturday 16 April.

Frank Hodges, the moderate Federation leader, hinted that the miners might be prepared to reach regional agreements, provided they were tied to the cost of living. The abandonment of a national scale was anathema to the rest of the union executives, however, and they rejected the plan. At this point, the railwaymen's union leader, J.H. Thomas, decided that the miners were being unreasonable and intransigent, and demanded that they went back to the negotiations. When they refused, he and the transport workers called off their strike on Friday 15 April. The miners were on their own again. The Triple Alliance was renamed the Cripple Alliance and the day went down in labour folklore as Black Friday. The result was that it turned out to be that old scenario after all; the miners were eventually forced to take severe pay cuts, in some regions as much as 40 per cent.

International politics, which had so far done very few favours for the mining industry, now unwittingly helped to produce a small recovery. The French moved into the Ruhr and the local German miners went on strike, providing a very welcome, if temporary, boost for coal exports. But attitudes in the industry were hardening, and in 1923 the federation got a new leader. Frank Hodges resigned and Arthur James Cook became the new secretary. A.J. Cook was born in 1883 in the Somerset village of Wookey, where he started working in the fields as a boy. In 1901, attracted by the higher wages on offer, he moved to the Rhondda. On his very first day underground a miner was killed in an accident near where he was working, and he helped carry the body back to the family. It made a very deep impression on him, as it would have done on any 18-year-old, but it also stirred him to thinking about the conditions in which men worked in this new way of life that he had just entered. He soon began taking an active interest in the union, and gradually began to move up the ladder. He had scarcely any formal schooling, but he was given a scholarship to the Central Labour College, where he discovered a new doctrine, Marxism: He was to become one of the first members of the British Communist Party, and though he didn't remain a member, he kept the same revolutionary outlook.

Cook became known as a brilliant orator, but his views were to bring him into conflict with authority. He spoke out against the First World War, and was soon being carefully watched by the police. In March 1918 he was charged with sedition and sent to gaol for three months. As one of the most passionate leaders of the Welsh miners during the 1921 lock-out, he was soon in trouble again. He was charged with incitement and unlawful assembly, and once again sent to prison, this time for two months' hard labour. It was an experience that coloured his later actions, and a humiliation he never forgot nor forgave: 'It is just six years since they not only handcuffed me but led me in chains from one end of the train, in Swansea station, to the other, in full view of the public. The same at Cardiff station.'[1] He brought a new militancy to the miners and caused consternation among the moderate leaders of the TUC. Fred Bramley, general secretary of the TUC, described him as 'a raving, tearing Communist' and prophesied that the miners were in for a hard time. They were, but not perhaps quite in the way that Bramley meant.

In 1925, Winston Churchill was appointed Chancellor of the Exchequer and in his first budget he announced that Britain was going to go back on the gold standard that the country had been forced to abandon in 1914. It was excellent news for the big bankers of the city, as it brought London back into its old position as a leading centre for world finance. It was equally welcomed by the patriotic members of the Conservative Party, who thought national dignity had been restored. It was disastrous for industry. The pound was now grossly overvalued, so the price of British exports rose in comparison with those of their competitors, while the cost of essential imports also became more expensive. The mining industry was immediately affected, as the combination of the return to the gold standard and a resumption of work in the Ruhr led to a 20-million-ton drop in coal exports and rising unemployment. The mine owners responded as they had in the past: they declared an end to the scheme that linked wages to profits, an end to national agreements and also announced wage cuts throughout the country. Some areas got away with small reductions, while other less profitable regions were severely hit. In the Forest of Dean, a married miner with three children would now be getting 4s 4d a day. If he had been on the dole he would have been receiving 4s 11½d a day. It was absurd that a working man would be better off unemployed, but if he gave up his job voluntarily he wasn't eligible for the dole. The miners' response was as predictable as the owners': they rejected the proposals and refused to negotiate any sort of deal unless the wage cuts were withdrawn.

The federation was not in a very strong position as there were large stockpiles of coal in the country, but when they were immediately promised support from other unions the government became seriously worried. They decided to buy time, by agreeing to subsidise wages at the current level for nine months, during which period there would be an official inquiry into how the mining industry could be returned to profit. The news was passed to the miners on Friday 31 July; it was marked down as 'Red Friday', a day to set against the earlier Black Friday. No one was under any illusion that the basic problems had been solved. When Cook remarked to Churchill that he was glad it had been settled, Churchill replied, 'You have done it over my blood-stained corpse'. There was a feeling that there would be trouble ahead once the agreement came to an end. In August, Herbert Smith, the president of the federation, addressed the Miners' Delegate Conference: 'Thirty-four weeks to go – Thirty-four weeks to go to what? To the termination of the mining agreement and the greatest struggle in the history of the British working class.' His forecast was all too accurate. In the meantime, the royal commission set about its task of analysing the woes of the industry and looking for solutions. Unlike earlier inquiries, neither miners nor owners were represented on the commission.

The basic facts that the commission came up with were incontrovertible: mines were losing money; more coal was being produced than could be sold, yet there were more men employed than ever before. The whole industry was in desperate need of reorganisation and modernising. It did not actually spell out the other factor in words, though individual members of the commission made their views plain: the two sides of the industry saw each other as implacable enemies, rather than parties working together for the good of the whole. When the report appeared, it pleased no one. It declared that a reduction in wages was inevitable, but this sacrifice by the men would be rewarded by great changes and improvements at some time in the not too distant future. It rejected the owners' claim that the men should work longer hours – not perhaps the most sensible suggestion when one of the major problems was that too much coal was already being produced. The commission also wanted greater controls of the industry, and the introduction at some time of such 'luxuries' as pithead baths and paid holidays.

There followed weeks of talks, during which Cook indicated to Prime Minister Stanley Baldwin that he might be prepared to accept the report if the government would guarantee that there would be a meaningful reorganisation. But Baldwin would not agree; he insisted that reorganisation should be left to negotiations between the owners and the federation. As the owners and the federation rarely managed to agree anything, this seemed very much like an idea with nowhere to go. There were more meetings. The unions wanted to put reorganisation at the top of the agenda; the owners insisted only on wage cuts, longer hours and a return to regional pay deals being discussed. It was a total impasse. The miners were no longer in any mood to compromise, and Cook addressed meetings all over the country. Arthur Horner, who often spoke with him on the platform, mused on why his own reasoned arguments received polite applause, but Cook would 'electrify the meeting'. Eventually the penny dropped: 'I was speaking *to* the meeting. Cook was speaking *for* the meeting. He was expressing the thoughts of his audience, I was trying to persuade them. He was the burning expression of their anger at the iniquities they were suffering.'[2] Ultimately, this was the true cause of the conflict: a confrontation that was following years of similar confrontations, few of which had been settled amicably, most ending in bitterness. In the end it was all expressed in one simple slogan: 'not a penny off the pay, not a second on the day.'

The result was inevitable. When the agreement ended on 1 May, the lock-out began and the General Council of the TUC called a special meeting to ask for support for the miners. One by one, the different unions backed the call, starting rather ominously, if alphabetically, with the Asylum Workers. The unions had had plenty of time to prepare for the strike, but so had the government. A special committee had been formed to consider how to deal with the crisis, which included setting up an organisation to recruit volunteers to keep essential services running. The strike began on 4 May, and is generally known as the General Strike, which suggests that every trade unionist in the country downed tools and stopped work. They didn't – nor were they expected to – but the essential feature for the miners was that virtually all transport workers supported them. The volunteers were brought in, mostly students, who rather enjoyed driving trains and buses, but it was never going to be a substitute for a properly organised transport system. Troops and the police were called in to stand by and wait for trouble, but the atmosphere remained calm in many areas. In Plymouth it was so peaceful that the police and strikers became bored with standing around and they organised a football match between themselves. In mining districts such as Newcastle, things were a good deal tenser. The print workers had been called out, and when some newspapers were produced, the strikers attacked the delivery vans and tried to destroy all the papers. The police charged in, and some leaders were arrested. Much more ominously, an attempt was made on 10 May to wreck the *Flying Scotsman* express, with 300 passengers on board.

Sir Herbert Samuel, who had chaired the 1925 royal commission, drew up a document based on the various points that had already been brought up in his earlier report. This was presented to the TUC negotiating committee, who happily accepted it as a basis for negotiation, even though none of the miners' delegates were actually at the meeting. There was a general sense of relief that the confrontation could be declared at an end, and the General Strike was called off. It was hailed as a victory for common sense, yet it gave the miners nothing and they refused to accept the deal. To them, it was Black Friday all over again, and they felt betrayed. Once more, being on their own and with no real funds to carry on the fight, the end was inevitable. Even though many remained on strike right through to the end of November, it was ultimately very clear who had won. Lord Birkenhead wrote in triumph to Lord Halifax: 'The discredit of the Miners' Federation is now complete. Torn by internal dissension they have been unable to prevent what are practically unfettered separate negotiations in each district.'[3] As he wrote, the government had come away with the trophy. The miners were to find many pennies off the pay, and a lot more than seconds on the day.

The victory did nothing to tackle the real issues that affected the industry. There was a woeful lack of modernisation. There were no real attempts to carry out research on new or more efficient ways of using coal. There were too many small pits that were unlikely ever to show a profit. Part of the problem was that even sober experts saw a future in which the demand for coal would go on increasing, even if in the 1920s there was a temporary blip. Professor Jarvis produced estimates based on a combination of home consumption and exports that began in 1911 with a total of 286 million tons, but which a century later would reach 784 million tons.[4] What was there to worry about? There was another problem faced by the miners when they tried to rouse public opinion. Simply because they lived in such closed, self-contained communities, very few people knew much about them, their way of life or the conditions under which they worked.

Very little had been written about miners in the nineteenth century, and what had been was never very flattering. Even George Eliot, who produced a sympathetic portrayal of a working-class reformer in the eponymous hero of *Felix Holt the Radical*, depicts the local miners as ignorant men, eager to riot and easily led by any smooth-tongued politician. There is nothing in English literature in that century to match the realistic depiction of the nature and hardships of the mining life that Zola described so vividly in his great novel *Germinal*. But in the twentieth century new writers came forward with very different attitudes, describing the mining districts and the people who lived in them in both fiction and non-fiction.

D.H. Lawrence was the first successful novelist to write about miners from personal experience. He was born in Eastwood in Nottinghamshire, near the Derbyshire border, in 1885. His father was a miner, a butty at the head of a small group of men. He received the pay for the whole group, which was then divided up either in the butty's own home or at one of the local pubs. Young Lawrence often had to go to the colliery office to collect the pay, and it is one of the memories that he used in his semi-autobiographical novel, *Sons and Lovers*. He described the acute embarrassment of his alter ego, Paul Morel, when, having collected the money for the men, he had to pay over the 'stoppings' for tools and rent:

> 'Sixteen an' six', said Mr Winterbottom.
>   The lad was too much upset to count. He pushed forward some loose silver and half a sovereign.
>   'How much do you think you've given me?' asked Mr Winterbottom.
>   The lad looked at him, but said nothing. He had not the faintest notion.
>   'Haven't you got a tongue in your head?'
>   Paul bit his lip, and pushed forward some more silver.
>   'Don't they teach you to count at the Board-School?' he asked.
>   'Nowt but Algebra and French,' said a collier.
>   'An' cheek an' 'impidence,' said another.

Lawrence was a sensitive boy, who hated the mining world. He had a very uneasy relationship with his rough father, who all too often came home drunk from the pub. Lawrence never made the miners out to be heroic figures, but he did appreciate that what they were was a product of the work they did and the conditions they endured. He did his best to be fair and there are moments in the novel when the father's more decent side shines through, as when he tells the children tales about his working life, finding a mouse running up the sleeve of his coat or describing the pit pony, Taffy:

> 'He's a brown 'un,' he would answer, 'an' not very high. Well, he comes i' th' stall wi' a rattle, an' then yo' 'ear 'im sneeze. "'Ello, Taff", you say, "what art sneezin' for? Bin ta'ein some snuff?"'
>   'An' 'e sneezes again. Then he slives up an' shoves 'is 'ead on yew, that cadin'.'
>   '"What's want, Taff" yo' say.'
>   'And what does he?' Arthur always asked.
>   'He wants a bit of bacca.'

The father's story of the pit pony fits in comfortably with what we know of real miners' attitudes to the hard-working animals, which often seemed to have personalities of their own. One pony was famous for having a fixed number of tubs that it would pull. Full or empty, it made no difference; the animal seemed to be able to count the number in the train and if there was one more than it reckoned to be its quota, it refused to budge an inch. The ponies also got incorporated into the men's list of superstitions and omens. Many miners believed that if a pony was fit and sprightly the day would go well, but if it was dull or restless then there was trouble ahead for the men working with that particular animal. The belief was so strong among some miners that they would refuse to work if their pony was ill, convinced that they would be caught in a rock fall or injured in one of the many other ways that a man could get hurt down a pit.

Every miner expected some sort of injury during his working life. Lawrence wrote about this and what it meant to the women of the pit villages:

> Morel was a heedless man, careless of danger. So he had endless accidents. Now, when Mrs. Morel heard the rattle of an empty coal-cart cease at her entry-end, she ran into the parlour to look, expecting almost to see her husband seated in the waggon, his face grey under his dirt, his body limp and sick with some hurt or other.

It was an experience common to all miners' wives, and when Morel had a more serious accident, a crushed leg that meant he had to go to hospital, his wife railed against the lack of facilities: 'You'd think they'd have a hospital here. The men bought the ground, and, my sirs, there'd be accidents enough to keep it going. But no, they must trail him ten miles in a slow ambulance to Nottingham. It's a crying shame!'

Much of what Lawrence wrote of the mining life rings true, and is clearly coloured by personal experience. The older Morel is not an admirable character and nor by all accounts was Lawrence's own father, but the author's real hatred is levelled against the system he had to live under and what he saw as the soul-destroying ugliness of the surroundings. Lawrence described The Bottoms, where the Morels lived, as 'substantial and very decent'. Anyone walking round the terraces saw little front gardens, planted with flowers under the parlour windows. But the parlours were hardly ever used. Life went on at the back of the house, in the kitchen, beyond which a back yard opened on to an alley between the houses, full of ash pits.

The kitchen was the centre of all the life of the house, and that life was tied to the working hours of the mine. For men working on an early shift, preparations began the previous night. The wife would have his food prepared – his snap – and his pit-flask would be filled, usually with tea. Cold tea was always considered the ideal drink down the mine. The fire was banked up. Like an Olympic flame it was never allowed to go out, and was either flaming brightly or glowing gently in its shiny, blackened grate. It was the fire that dried out the mine-drenched pit clothes and heated the water for the kitchen tub, where the man of the house washed away the coal dust. A familiar part of the housewife's routine was scrubbing her husband's back. When the men left for work in the morning, the village was left to the women and children. There were chores to be done, but there was also time to gossip over the back walls, while the youngest children played among the ashes of the back alleys. The routine only changed on paydays when, after the money had been shared out, the men headed for the pub and the women to the market.

Time and again, Lawrence balances any sympathetic remarks about the miners and their lives with tirades against what he sees as the ugliness of the surroundings and the boorishness of the men. He was not alone in his views. The 1920s and 1930s were a great age for documentary observation. Filmmakers such as John Grierson looked at the working lives of the ordinary people of Britain, and novelists such as George Orwell and J.B. Priestley went off to investigate parts of Britain that seemed to them to be foreign countries. One of the best known examples is Orwell's *The Road to Wigan Pier*, but Priestley also had some telling words about the lives of the Durham miners that were very similar to Lawrence's ideas:

> The time he does not spend underground is spent in towns and villages that are monuments of mean ugliness. Some people shall tell me that this does not matter because miners, never having known anything else, are entirely indifferent and impervious to such ugliness. I believe this view to be as false as it is mean, Miners and their wives and children are not members of some troglodyte race but ordinary human beings, and as such are partly at the mercy of their surroundings. I do not want to pretend that they are wincing aesthetes, unnerved by certain shades of green and sent into ecstasies by one particular pink. Their environment must bring them to despair – as I know from my own experience that it frequently does – or in the end it must blunt their senses and taste, harden the feelings and cloud the mind. And the latter is a tragic process, which nothing that calls itself a democratic civilisation has any right to encourage.[5]

It was not just a general ugliness that Priestley saw, there were some places that seemed so strange that, as he said, if he had not been standing there, he would have thought of them as fantasies – 'a symbol of greedy, careless, cynical, barbaric industrialism'. Nowhere made a deeper impression than the Durham village of Shotton and its immense tip:

> Imagine then a village consisting of a few shops, a public-house, and a clutter of dirty little houses, all at the base of what looked at first like an active volcano. This volcano was the notorious Shotton 'tip', literally a man-made smoking hill. From its peak ran a colossal aerial flight to the pithead far below. It had a few satellite pyramids, mere dwarfs compared with this giant; and down one of them

a very dirty little boy was tobogganing. The 'tip' itself towered to the sky and its vast dark bulk, steaming and smoking at various levels, blotted out all the landscape at the back of the village. Its lowest slope was only a few yards from the miserable cluster of houses. One seemed to be looking at a Gibraltar made of coal dust and slag. But it was not merely a matter of sight. That monster was not smoking there for nothing. The atmosphere with thickened with ashes and sulphuric fumes; like that of Pompeii, as we are told, on the eve of its destruction. I do not mean that by standing in one particular place you could find traces of ash in the air and could detect a whiff of sulphur. I mean that the whole village and everybody in it was buried in this thick reek, was smothered in ashes and sulphuric fumes. Wherever I stood they made me gasp and cough.

It certainly seems that writers such as Priestley were looking to paint the blackest picture that they could, probably because of the shock at what they found in these 'foreign' territories. On the other hand, having painted his gloomy landscape, he then proposed a tongue-in-cheek political remedy:

Suppose we had a government that began announcing; 'Coal is a national necessity. But it is unjust that one class of men should do this hard, dangerous work at starvation wages. During the war we had conscription. We will now have conscription again, this time for the coal mines, where every able-bodied man shall take his turn, at the usual rates of pay. All men in the Mayfair, Belgravia, Bayswater and Kensington areas who have received Form 5673D will report themselves at King's Cross and St. Pancras stations on Tuesday next, for colliery duty.' What a glorious shindig there would be then! And if you could buy yourself out by subsidising a professional miner, how the wages in East Durham would rise!

> Grimethorpe Colliery Band was formed in 1917. Just five days before they were due to appear at the National Brass Band Competition in 1992, the closure of the colliery was announced. They won, and the band continues to play with the same name, continuing a fine tradition.

What one seldom gets in any of these accounts is any sense that the miners and their families ever had any pleasures in life. Priestley did write about the men's love of gambling, and Lawrence certainly went on at great length about the visits to the pub, but there was another side of the story that seldom got told. Sport played a part in many colliery districts. Rather surprisingly, one of the sports that became very popular in some areas was the one that had its origins in a very different world, in the famous public school of Rugby. Rugby Union reached South Wales in the 1890s and took a hold that has never been relaxed. A miner from the Rhondda remembered his school days in the 1950s when any boys caught kicking a round ball in playtime were given a clip round the ear by the sports master. The game also became popular in the north of England, but the players had a dilemma. They were attracting large crowds but they got no pay for taking time off work, and if they were injured in a game where injuries are not that uncommon, they got no compensation. In the 1890s some northern clubs began to pay their players, which led to a rift with the official body, mainly made up of the wealthy southern clubs, voting against allowing any payments. In 1895 many of the clubs in Lancashire and Yorkshire broke away and eventually formed themselves into a new body and changed the rules of the game, becoming Rugby League. In the summer months, cricket was popular in northern England, and it used to be said that if Yorkshire needed a new terrifyingly fast bowler, all they needed to do was whistle down the nearest pit and one would emerge.

There was another aspect to mining life that Lawrence does mention. Miners were not really urban workers in the way that factory workers were. The countryside was never far away. Even the Black Country had yet to be absorbed into the great conurbation that was spreading out from Birmingham. In the valleys of South Wales, the sheep from the hill farms could wander down to the mining villages in the valleys. This landscape offered some respite, a contrast to the dark, cramped tunnels in which the men spent so much of their lives. However, as the twenties gave way to the thirties there was little that brought any comfort to the working people of Britain, as the country headed into the great Depression.

CHAPTER FOURTEEN

DEPRESSION AND WAR

Immediately following the end of the General Strike, the government came up with the Trade Disputes and Trade Union Act, which made illegal any strike that 'has any object other than or in addition to the furtherance of a trade dispute within the trade or industry in which the strikes are engaged'. In other words, there would be no more alliances, triple or otherwise. It also made it illegal to go on strike 'to coerce the government either directly or by inflicting hardship upon the community'. There were other clauses making intimidation illegal, and setting out what intimidation might consist of. The act further tried to limit the influence of the Labour Party by changing the rules on the political levy paid by union members: now members had to opt to pay the levy rather than opt out. It was resented by the miners' union on principle, but made little practical difference in the short term. Their recent experience of relying on sympathetic strikes had hardly encouraged them to look for new alliances. The effect on the Labour Party was also rather less than they feared: they formed a government in 1928 under Ramsay MacDonald.

When the Labour Party took office, they did so with some optimism, as the economy seemed to be going through a period of slow but steady growth. MacDonald declared that one of his main objectives would be to reduce unemployment. That sense of optimism was ended on just one day, Thursday 24 October 1929, the day that Wall Street collapsed, wiping $16 million off securities and sending many companies into bankruptcy. Its effects were felt around the world, and any hopes of reducing unemployment vanished as Britain slid into a deep depression. The unemployment figures grew steadily worse, until by August 1930 they had passed the 2 million mark. MacDonald still wanted to do something to help the miners. The result was the passing of the Coal Mines Act of 1930, which set up a system of quotas. A council was appointed to decide how much coal was to be raised each year, and then allocated the permitted output for each pit. An unfortunate side effect of the measure that was supposed to bring order to the industry was that it passed out the quotas without any regard to the efficiency of the different collieries. The modernised, efficient collieries were forced to produce less than they wanted, while the small, inefficient concerns could work at more or less their maximum capacity – some even managed to sell on the rights to their quotas to bigger pits. It was supposed to help keep miners at work; it failed. By 1932, 40 per cent of miners were out of a job.

There was a second part to the 1930 act, which called for the setting up of a Coal Mines Reorganisation Committee that was supposed to promote the amalgamation of collieries to form more efficient units. It was a toothless piece of legislation with no power of enforcement, and as a result the coal owners simply ignored it. Six years later, not a single example of voluntary amalgamation could be reported in any of the coalfields. The act that had seemed to promise much had delivered little.

For many miners, all they had was the grim reality of being without work or any prospect of work. As long as they had insurance stamps, they received a fixed amount of unemployment pay. As soon as the stamps ran out there was the dole, but that was only provided for those who could prove they needed the money – not having a job was not enough. The unemployed were subjected to a means test that was based on the earnings for the whole family. Everything was counted: from money from insurance policies to savings and earnings from any member of the household. It seemed that the official policy was not to help the unemployed poor, but to minimise the payments that had to be made. Cases were reported of homes where a grandmother or grandfather lived with the family, and the dole was reduced by the amount the officials deemed they should have been paying in rent, regardless of whether they were actually paying anything or not. A miner whose sons or daughters had jobs could find himself with nothing. This caused bitter resentment, not just among the older generation but also among the young, who suddenly found that the little they earned was now expected to keep their parents as well as themselves. As a result, many of the young moved into lodgings, but even then if they were seen visiting their parents they could be reported by unkind neighbours and the money was docked.

Drawing the dole was a humiliation made worse by the bureaucracy that surrounded it. People loathed the means test, and they were right to do so. Deprived of a reasonable income, families were often reduced to desperate measures: scrabbling over old spoil heaps to find nuggets of coal to keep their houses warm, or even just to provide heat to cook their meals. George Orwell gave a vivid description of what it all meant to the families:

> You can't get much meat for threepence, but you can get a lot of fish-and-chips. Milk costs threepence a pint, and even 'mild' beer costs fourpence, but aspirins are seven a penny and you can wring forty cups of tea out of a quarter-pound packet. And above all there is gambling, the cheapest of all luxuries. Even people

on the verge of starvation can buy a few days' hope ('Something to live for,' as they call it) by having a penny on a sweepstake.[1]

For many families in the mining communities there was no hope of getting any sort of job. Wales was particularly badly hit. Official statistics produced by the Ministry of Labour for Merthyr Tydfil in 1934 show that 61.9 per cent of workers were unemployed. Certain areas of Britain were officially designated as 'depressed' and unpaid commissioners were appointed to do something about it. In 1934 Parliament made £2 million available to deal with unemployment in four areas: South Wales, West Cumberland, Tyneside and Scotland. The act was heavily criticised as being hopelessly inadequate, and inevitably comparisons were made with the billions being poured into work creation schemes in Roosevelt's New Deal on the other side of the Atlantic. Efforts to persuade industries to move into the depressed areas were largely unsuccessful. Sir Malcolm Stewart, the first commissioner appointed for England and Wales, had approached nearly 6,000 firms, with an almost total lack of success. The most striking demonstration of the misery created during the slump was the hunger march of the men of Jarrow. This was a Tyneside shipbuilding community that had fallen victim to a scheme devised by the successful shipbuilder Sir James Lithgow. He formed a consortium with other successful owners to buy up less successful yards and close them down, with a guarantee that they would not reopen for at least forty years. Palmers of Jarrow was one of them, and virtually the only employer of any size in the town. As a result, Jarrow was in the unhappy position of having even worse unemployment figures than Merthyr, with three out of four unemployed. The march of the men from Jarrow to London received huge publicity and it was widely recognised that they were not just marching for the families of their own impoverished town, but that they symbolised the needs of the unemployed everywhere.

In the late 1930s, there were improvements in the coal industry: annual output that had dropped by 40 million tons between 1930 and 1934 was back at the 1930 level by 1937; unemployment in the industry that had stood at over 40 per cent in 1932 was down to 19 per cent by 1937. The working miners were even able to get a modest shilling a shift wage rise in 1936. However, in other respects very little had changed. Mechanisation developed slowly and was virtually unknown in many collieries. All attempts at amalgamations had failed, and there remained a multitude of small pits where a nineteenth-century miner brought back to life would find little had changed. Even the most modern pits could not compare with those in America, for example, where working methods had been transformed. In the past when mines were faced by growing competition from overseas the first response had been to reduce their costs by cutting wages, instead of investing in new technology. If the owners had been unwilling to put money into improvements when the economy was reasonably healthy, there was no way they were going to do so in the middle of an economic crisis. The life of the miners also remained largely unchanged. For every miner who ended his shift at pithead baths and came home in clean clothes there were hundreds who still left work covered in the dirt of the pit. For every house with its own bathroom there were rows of terraces where the men sat in zinc tubs in front of the kitchen range.

Although mechanisation was slow, it was happening in some regions. At the beginning of the century only 1.5 per cent of British coal was cut by machine, but the proportion did steadily increase. Cutters were improved by the introduction of new alloy steels that made them far more efficient. In 1902, W.C. Blackett invented an endless steel conveyor belt that was installed at a Durham colliery, and three years later Richard Sutcliffe introduced his version of the endless conveyor to a Yorkshire pit. This was clearly more efficient than filling tubs, but it did not necessarily make working life much easier. George Orwell went down a Lancashire mine, and was taken to the coalface, roughly 3ft high:

> The first impression of all, overmastering everything else for a while, is the frightful, deafening din from the conveyor belt which carries the coal away. You cannot see very far, because the fog of coal dust throws back the beam of your lamp, but you can see on either side of you the line of half-naked kneeling men, one to every four or five yards, driving their shovels under the fallen coal and flinging it swiftly over their left shoulders. They are feeding it on to the conveyor belt, a moving rubber belt a couple of feet wide which runs a yard or two behind them. Down this belt a glittering river of coal races constantly … It is a dreadful job that they do, an almost superhuman job by the standards of an ordinary person. For they are not only shifting monstrous quantities of coal, they are also doing it in a position that doubles or trebles the work. They have got to remain kneeling all the while – they could hardly rise from their knees without hitting the ceiling – and you can easily see by trying it what a tremendous effort this means. Shovelling is comparatively easy when you are standing up, because you can use your knee and thigh to drive the shovel along; kneeling down, the whole of the strain is thrown upon your arm and belly muscles. And the other conditions do not exactly make things easier. There is the heat – it varies, but in some mines it is suffocating – and the coal dust that stuffs up your throat and nostrils and collects along your eyelids, and the unending rattle of the conveyor belt, which in that confined space is rather like the rattle of a machine-gun.[2]

Machinery brought a new three-day working system. The machine-cut coal was loaded on to the conveyor by one day shift. The next two days were taken up with cutting more coal, putting in supports and moving the conveyor forward. But throughout the thirties, there were still far too many collieries where mining methods had remained unchanged for decades.

There was one final attempt to impose change on the industry. The Labour government had gone, replaced by a new administration, a National

Government under Neville Chamberlain. In 1938 they proposed strengthening the commission that would impose amalgamations, but when the owners objected they meekly abandoned the plan. They did, however, pass one surprising piece of legislation: they nationalised mine royalties, the payments made to the owners of the land. The owners received compensation, and where they would not agree to the assessment it went to arbitration. The original government offer, based on an estimate of fifteen years' income, had been for total payments of £75 million, against the owners' own estimate of £112 million. The owners insisted that the offer was sent to arbitration, and soon regretted it. When the arbitration was complete they finished off in a worse position than when they had started: the award was for £66.5 million. Happily for them, the government was very much on the mine owners' side and promptly coughed up an extra £10 million. It is worth noting at this point that the government had considered £2 million a generous amount to pay to help find work for millions of unemployed workers.

Unemployment stopped becoming an issue in 1939 when Britain went to war with Germany. Surprisingly, coal mining was not listed as a reserved occupation and the younger miners were soon being called up to join the armed services. There was also a movement away from mining to the new, better paid occupations, such as working in munitions factories. Coal, however, was essential and eventually it became clear that there was a very real danger of a fuel shortage that would affect everything from the running of factories to the nation's transport system. By the summer of 1943 it was estimated that 36,000 men had left the industry and to get coal production up to the necessary level, 40,000 new workers would be needed to take their place. Ernest Bevin, the Minister of Labour and National Service, came up with a scheme for conscripting workers. Each month, ten numbers were placed in a hat and two of them were drawn out: the conscripts whose call-up papers ended in those two numbers were sent to the mines instead of into the army, navy or air force. There was also a possibility for conscripts to opt for the mines. Whether sent to the mines by the luck of the draw or from choice, they were all known as the 'Bevin Boys'.

Altogether, 48,000 Bevin Boys worked in the mines. Many of them would far rather have been in the armed forces. A website now records the words of some of those who were, you could say, sent down when they were called up.[3] One 17-year-old had been in the Air Training Corps in Devon and

Around 48,000 young men and boys were conscripted to work down the mines as Bevin Boys during the Second World War. Unlike conscripts into the armed services, they received no official recognition until 2008, when survivors were awarded a special Veteran's Badge.

volunteered for the Fleet Air Arm – but was sent to a colliery in Pontefract. He was very unhappy with the decision, and his misery was made even worse when his train, instead of arriving at the scheduled time of 8.30 p.m., rolled into Pontefract station at 2.30 a.m. He spent his first night as a miner sleeping in the only bed he could find, in a police cell. Others had similar tales of having been in cadet forces and of actually looking forward to service life, and instead of using their experience, found themselves in an alien world of which they knew nothing. To make matters worse, because they had no uniforms or special insignia, they were often taunted with cowardice, or accused of being conscientious objectors or even deserters. Some found it especially hard. One boy suffered from claustrophobia, and was sent for psychiatric assessment. He did not receive a sympathetic hearing, being told to go down the pit or spend three months in gaol. By April 1944 *The Colliery Guardian* reported that 135 of the boys had been prosecuted for failing to obey their call-up orders. Thirty-two of them went to prison, but nineteen rapidly changed their minds, were released and went to work.

The Bevin Boys, unlike the regular miners, were given work clothes, safety helmets and work boots, but they had to pay for their own tools. As many of them pointed out, if they had been called up for the army no one would have expected them to pay for their own rifles. If they resented the treatment they were receiving, they were also resented by many in the mining community. Families could not understand why their sons, who were brought up to the mining life, were being sent off to war, while strangers were being brought in to take their place. Propaganda films of the time showed the Bevin Boys in a more favourable life. One of them concentrated on two very middle-class boys who, when they weren't below ground, seemed to be having an idyllic time wandering through woodland studying for their university entrance exams. In fact, the only black spot in this sunny existence was that one of the boys was handicapped in his music studies because the miner's family they were staying with didn't have a piano.

In practice, unless they had special technical skills, as electricians or engineers for example, the Boys played a secondary role underground. Very few became face workers, hewing the coal. Most worked as assistants, filling tubs and doing other, less skilled jobs. The conditions they found varied enormously, as did the reaction of the men they had to work with. One of the Boys heard a curious gurgling noise coming from a hole that had been drilled for shot firing. He asked what it was. The men were used to the noise of water, released from the rock, but they told him it was a 'coal frog'. He peered into the hole, the men roared with laughter and for the rest of his time down the mine he was known as 'Coal Frog'. There has always been a place for humour down the mine, and there must have been a good deal of laughter when young Eric Bartholomew arrived for work. He was released after just thirteen months because of a heart complaint. After the war, he went into showbusiness and changed his name to Eric Morecambe. For some it was an experience that they looked back on with considerable pride. Others were only too glad to leave, and they did so with little or no thanks from anyone. Whilst others who had served their country got demob suits, service medals and a guarantee of a return to their earlier occupations, the Bevin Boys got nothing and some were not released until 1948.

The Bevin Boys were shabbily treated. They had done their bit, and many paid a high price, suffering in later life from pneumoconiosis, yet they had to wait until 1995 before they were allowed to join other war veterans at the annual Remembrance Day service in London. In 2008 they finally received badges commemorating their work.

The end of the war in Europe in May 1945 brought a General Election, with a landslide victory for Labour. It also brought hope to the mining communities that they might actually see real changes in their industry. Other mining areas had no such optimistic outlook. In the south-west, following the collapse of the copper industry in the late nineteenth century, some mines survived and even prospered by turning to tin mining. There was still a high demand for the metal, particularly from the tin-plating industry – this was a time when tin cans were really made from tin-coated iron. But by the 1930s, that industry too was in decline, unable to compete with foreign imports. The last tin-smelting works in Cornwall closed in 1931. However, when the war began, foreign imports no longer looked so attractive. The Cornish Tin Mining Advisory Committee was set up by the Ministry of Supply to investigate how supplies could be increased. A.K. Hamilton Jenkin was one of the committee members and he was not impressed:

> I listened to our terms of reference – we had twelve months at most to get into production, our labour supply was to be scanty, and as for machinery, we must make do and mend. At best under the circumstances we could only advise the reopening of a few old shallow properties, and the testing of 'dumps' which had in fact been turned over many times before for such mineral as they contained. We had the money but not the time to reach the main ore bodies, even though in Cornwall these are far from deep according to modern mining standards. Under the circumstances it is easily understandable if the results of this hasty war-time activity were meagre.[4]

The only good that came out of the scheme was that the re-examination of the geological records had revealed 'great scope for development'. So there were some grounds for optimism, but many miners had already left the industry, finding new employment in the far more profitable china clay workings, and there was little enthusiasm among investors for tin and copper mining. Whatever the prospects might be for coal, there was little hope of anything much being done in any of the metal-mining industries.

CHAPTER FIFTEEN

COAL FOR THE NATION

Nationalisation of the coal industry had long been the ultimate dream of many miners and now that it was about to come into being, the big question remained: what form would it take? The Cabinet discussed the issues in December 1945.[1] The main aims were succinctly expressed: 'The Bill was based on the assumption that the industry must pay its own way. In order to do so it must meet its current expenses and interest and amortisation on the compensation payments.' A warning note was sounded by Aneurin Bevan, the Health Secretary, born in Tredegar in South Wales. He had started his working life as a miner, and he had no illusions about what that life entailed:

> Here down below are the sudden perils – runaway trams hurtling down the lines; frightened ponies kicking and mauling in the dark, explosions, fire, drowning. And if he escapes? There is a tiredness which comes as a reward of exertion, a matter of relaxed limbs and muscles. And there is a tiredness which leads to stupor, which remains with you on getting up, and which forms a dull, persistent background to your consciousness. This is the tiredness of the miner, particularly of the boy of fourteen or fifteen who falls asleep over his meals and wakes up hours later to find his evening has gone and there is nothing before him but bed and another day's wrestling with inert matter.[2]

Bevan was the one Cabinet member with first-hand experience of working below ground, and he knew what the men felt and what they expected:

> One of the problems of the industry was to remove the feeling among the workers that they were working to make profits for the owners, and, while he did not object to the substance of the proposals, he suggested that, for psychological reasons, it was important to avoid any suggestion that under the new administration miners would still be working to provide payment to the owners for their former profits.

A few days later, the Minister of Fuel and Power, Emmanuel Shinwell, presented the results of his discussions with the miners' union, now re-formed as the National Union of Mineworkers (NUM). He made the priorities clear: to increase coal production and to recruit new workers to the industry:

> The National Union of Mineworkers from the outset promised collaboration in increasing production and set up an organisation under Mr. Horner [Arthur Horner, General Secretary of the NUM] for this purpose. They also undertook to deal with absenteeism through the Lodges following my decision to discontinue prosecutions. Their efforts have met with some success in certain districts and lodges, but are far from having been universally successful either in increasing production or in dealing with absenteeism. As regards recruitment, certain miners' leaders have found it difficult to discontinue old habits of disparaging the industry in the mind of the public by dwelling on the arduous and dangerous nature of coalmining and the unwillingness of miners' sons to enter it.[3]

It would perhaps be more than a little naive to believe that an act of Parliament would at a stroke annihilate a deep-rooted bitterness born of generations of conflict. Shinwell also noted that the miners were expecting something more than a mere change of ownership. They were looking for better working conditions: a five-day week and two weeks' paid holiday instead of just one, together with improved compensation for death and injuries. He was sympathetic, but was doing no more than promising to consider the issues compassionately.

Whatever the discussions that were taking place in the Cabinet Office, the miners of Britain were full of optimism. Vesting Day was 1 January 1947, which was celebrated by miners throughout the country. The flag of the new National Coal Board (NCB) was hoisted in front of cheering crowds and miners fastened name boards to colliery walls proclaiming a new era: 'This colliery is now managed by the National Coal Board for the people.' Many thought that the change would bring in a wholly new attitude to how the industry would be run, how prices would be set and how pay levels would be judged. However, as the Cabinet discussions made clear, the NCB was required to make a profit and, not only that, it also had to make provision for other payments. Chief among these were paying interest to the government on money invested in the industry and putting aside funds for future investment, paying back the interest on the compensation paid to the former owners and setting up a fund that would eventually pay off those loans. The initial amount set aside as compensation for 'coal industry value' was £164.66 million, but allowances had been made for extra payments for unspecified assets. The eventual bill came to £388 million. This was a huge debt to hang round the neck of the infant NCB and, in spite of Nye Bevan's words of warning, it seemed that the miners would still be digging coal to pay their old masters for many years to come.

The NCB did what it could in difficult circumstances that affected not just the coal industry but the whole country, burdened by war debts. Modernisation moved forward and the welfare of the workforce improved. It had to, for there was an urgent need to recruit more workers, and very few were prepared to accept the conditions that had been taken as normal in far too many pre-war collieries. In particular, the young men, many of whom had just come out of the forces, were not prepared to put up with walking home in their pit clothes, taking the grime of the colliery back into their houses. Pithead baths had been provided in some collieries and there was a major push forward in the late 1930s using money from the Miners' Welfare Fund. This had been set up by a statute of 1920 that required mine owners to put in a penny for every ton of coal sold, and the coal owners to provide a shilling in the pound on their royalties. In 1937 over £600,000 was made available specifically to build baths. The process was continued after the war, and it made an enormous difference, quite as much to the wives as it did to the men. They no longer had filthy, wet clothes hanging around the house every single working day; they no longer had to scrub their men's backs in front of the kitchen fire or heat great pans of water by the kitchen range.

Under the new system, the men arrived at work in their everyday clothes, which they put in a locker, one of a range of 'clean' lockers. They took their soap and towel, went through the bath room, with its rows of showers, to the 'dirty' lockers. There they changed into their work clothes and set off to the pithead. At the end of the day, the whole process was reversed, with the obvious exception that this time they had a good, hot shower between the dirty and the clean areas. The nature of the work was also beginning to change.

Improvements had been forced on the industry during the war years as a result of the severe manpower shortage. Development of a machine that could both cut and load coal began in 1942 and ended with a successful trial of a moving coal cutter that dragged a loading unit behind it. In its final form, it consisted of an AB coal cutter mounted on a Meco-Moore loader. The machines could not be used on the old pillar-and-stall system, so that almost all coal was produced on longwall faces. The original conveyor belts, of the type described in the last chapter, were efficient up to a point, but needed to be well maintained. If repairs were neglected the results could be disastrous. At Creswell Colliery, Nottinghamshire, in 1950, an inspector reported that one of the belts was showing signs of wear. It was the evening of 25 September, and an order was given to make repairs, but was overruled because there was still coal to be shifted and it seemed to be working without any obvious problems. In the small hours of the following morning, the next shift discovered that in parts the belt was virtually shredded. The result was a huge increase in friction, causing the belt to get hotter and hotter until, before anything could be done, it caught fire. The first miners on the scene tried to treat the blaze with a pair of fire extinguishers, but they were hopelessly inadequate. The fire fighters were called in, but there was not enough pressure for their hoses underground, and the fire raged on. Rescue parties rushed to the scene, but the noxious carbon monoxide from the burning rubber and the heat drove them back. It was all too obvious that the men trapped behind the fire could not survive and, reluctantly, the rescue attempt was abandoned. Eighty men lost their lives in the disaster.

The Creswell disaster led to the development of fire-resistant belting. Further advances, based on the armoured conveyor developed in Germany, also helped to increase safety and speed up production. All this new machinery took up a lot of space, so there was a new need to provide better supports that could be put in place quickly as the face advanced. In response, the Dowty Mining Equipment Company developed powered hydraulic props in the 1950s that moved forward as the face advanced.

Going down a mine in the second half of the twentieth century was very different from what it had been even a few decades earlier. Once the miner had changed into his work clothes, he would go to the lamp room to collect a lamp to clip on to his helmet and a battery that was fastened to a belt. He also collected a respirator that would provide him with up to two hours of air to breathe after any accident that released gas. All men going down were given a quick check, rather like the frisking you normally expect only to see in crime movies, to make sure nothing inflammable was going down the pit. Each man was then handed a numbered token, to record that he was underground, and he would be assumed to be still down there until the token was returned. Then it was time to descend, and even that had changed. The old steam winders had been largely replaced by ac electric winders by the 1940s, and the new machines were able to take far heavier loads of up to 12 tons at a time. Underground, too, appearances had changed dramatically. Wide, main roadways, their rooves supported by steel arches, were now lit by electric light that not only made life more comfortable but also helped to eliminate a disease that had been prevalent among underground workers. Being forced to work by the dim illumination of Davy lamps, many men developed nystagmus, a disease that causes a flickering eye movement and badly reduced vision.

Getting from the bottom of the shaft to the face had earlier meant walking but, increasingly, other systems were used. The most sophisticated involved being taken in comfort in simple open carriages hauled by a battery-powered locomotive. A less comfortable alternative was to ride the conveyor belt, and an interesting experience that proves to be: lying prone on the belt as it passes over the rollers was rather like receiving a very rough massage. I rode one of these on a visit to a colliery once, and after it came to an end I had the usual stooped walk to contend with. It all seemed about as dangerous as a visit to the London Underground, until one of the miners pointed to the pipes drilled through the strata at regular intervals. These carry away several hundred thousand cubic feet of methane each and every working day. You are also reminded that coal dust is the other deadly killer as you tramp through powdered stone spread across the floor to help keep down the coal dust. There was a section where the roof was being repaired by men attacking the rock with picks, an uncomfortable reminder that although you might be in a neat, well-lined tunnel, there are thousands of tons of rock up there over your head.

Nearing the face, the comforting electric light came to an end and I was left with the helmet lamp to help me move forward to where men were operating the cutter. In front of them was a pile of shattered coal, over which I scrambled on hands and knees to reach the face itself, stretching away into the darkness as a black, shining mass of coal. Now the roof was only supported by the hydraulic props and seemed uncomfortably close to the ground. My guide moved with ease, and I followed as best I could with my helmet beating a staccato rhythm against the roof as I never quite managed to keep my bent posture.

The cutter worked away, throwing up plumes of water and coal dust that fell back as black rain. No pickaxes here, no shovels: the modern miner is as much a technician as a manual worker. A few of the men wore facemasks, but the heat was so intense that they soon become uncomfortable. They worked with machines just as men in factories tend machines, but in very different conditions. One of the men explained: 'If you're a fitter you expect to get your hands dirty. You'll be dirty up to here', indicating his elbow. 'In the pit you'll be filthy all over.' A mechanic has to make his way into a dark, confined space, lit only by his helmet light, and go to work on a machine covered in slime and coal dust. I was told that it was not uncommon for men to come to the NCB for the excellent training facilities, and then get out to use their skills in less demanding surroundings. As someone else remarked, there are only two things that would make you come down a pit: shorter hours or better pay than you can get on the surface: 'Anyone who comes down here because he likes it doesn't belong in a pit. He belongs in t'bloody loony bin.' The final word on this particular visit goes to one of the deputies: 'Visitors come round this pit and they see this face, and they say they wouldn't work here, not for a hundred pound a week. Not for a hundred pound. And these are the best conditions anywhere in the country. God knows what they'd say if they saw the others!'

This then was the new world of coal mining many years after nationalisation: technologically more advanced than it had ever been and with better facilities for the men than ever before, but still a harsh, unforgiving workplace. Why, indeed, would you choose to work there without an appropriate reward? On 1 January 1947 it must have seemed that if the miners were now going to be producing coal for the people, and not just to make a profit for mine owners, they might indeed get that long-awaited reward. They certainly did enough to earn it. Coal production in Britain had peaked in 1913 at 287 million tons, but by 1946 it was down to 181 million. Ten years after nationalisation and production was right back at 224 million tons with no increase in manpower, mainly due to mechanisation. Wages had doubled in the ten years, which sounds very good until you take inflation into account. Basic prices had risen by more than half. Even so, by the end of the 1950s the effects of modernisation were really being felt. Sadly for the industry, just as it was getting back to normality, the demand for coal began to fall. The men who had been exhorted to produce more coal were now being told to produce less; campaigners who had fought for a five-day week now had one imposed on them.

There was no single factor involved. For years London had been noted for its pea-soup fogs, which were caused by a mixture of weather conditions and the pall of smoke from millions of coal fires, that brought the fog a new name: 'smog'. One of the worst ever smogs blanketed the city in 1952. It caused huge disruption to transport, and by the time it finally cleared it was estimated that 20,000 people had died from respiratory complaints. Something had to be done, and the eventual remedy was the Clean Air Act of 1956, which designated many of Britain's cities and towns as smokeless zones. The days of the coal fire were numbered, as more and more people turned to central heating systems powered by electricity, oil or gas. The move to gas was not necessarily bad news for the coal industry at first, since gas could only be produced by burning coal. However, by the early 1960s the first natural gas was beginning to be produced from North Sea drilling and, one by one, the old gasworks closed down. There was more bad news on the transport front. The days of steam locomotives on the railways and steamers on the sea of the world were numbered – British Rail was to build its very last steam locomotive in 1958. Export markets had virtually disappeared. Two major consumers remained: the electric power industry that still relied mainly on coal-fired generators, and the coking works that supplied the steel industry. But it was obvious that this was not like other dips in demand – there were markets that were closed to coal and they would never reopen.

By the end of the 1950s, the miners were looking at an increasingly bleak future. It was not exactly like the 1930s where men had to fight for every penny they earned, but there was a gnawing uncertainty: will this colliery still be open next year? Will miners have jobs to go to? Between 1965 and 1971, 229 British pits were closed down. In the vast majority of the cases, the miners and the NUM cooperated. They accepted the argument that, faced with a shrinking market, the least efficient pits had to go. No one wanted to see miners lose their jobs, but there was no escaping economic realities. The union went even further. In a shrinking industry they might have been expected to oppose the introduction of new machinery that would inevitably mean yet more job losses, but they accepted that change was now essential. By the 1960s the message was clear: modernise or die.

Change was not just about new working methods. In 1966 the NUM agreed to operate the National Power Loading Agreement (NPLA) that introduced a system for which the unions had been arguing for decades: it replaced district bargaining with a national award. It brought about a levelling-out of pay rates, which was good news for those who got increases but less happy tidings for those who saw their wage packets reduced. It was accepted, however, that in the new mining world of machine cutting, there was no longer an obvious distinction to be made between the old elite, who worked at the coalface, and the rest of the men. All these changes that were carried forward with the full co-operation of the workforce produced dramatic results. Productivity figures rocketed, from 25cwt per man per shift in 1951 to 45cwt in 1970. This was a change from the old days: unions and management working together for the industry as a whole. There was a good deal of grumbling among left-wing factions that the union leader Joe Gormley had given away too much for too little in the way of concessions in return, but he had the support of a large majority of the workforce.

Between 1967 and 1971 productivity in the coal mining industry increased by 20 per cent, but wages dropped from being 107 per cent of the national industrial average to 93 per cent. The resulting strike led to a three-day week and brought down the Conservative government of Edward Heath.

The miners might quite reasonably have expected that the efforts they were making would eventually be rewarded, but instead of getting better pay they were steadily slipping down the industrial wages league table. In 1967 the miners were earning 107 per cent of the national average; by April 1971 that had slumped to 93 per cent. As the miner quoted in the last chapter had said – only an idiot would work down a mine unless he got better pay than he would from an easier job. The men were not idiots; they knew that over that period productivity had risen by 20 per cent while their relative income had been dropping. Those who could get away left, while those who remained began to speak more and more frequently about the need for a pay rise. The new age of co-operation looked to be giving way to the old days of confrontation.

The problems faced by the miners in putting forward their wage demands was that they were not really able to have a free argument with their immediate employers, the NCB. The latter were ultimately controlled by government policy, and the government had decreed a limit to wage rises. On 5 January 1972 the unions rejected the small pay rise offered by the NCB, at which point the NCB withdrew all offers and broke off negotiations. On 9 January the union called the first national coal strike since 1926, and every pit in the country was closed. The initial reaction was summed up in the *Financial Times* of that day: 'The overall position of the miners is not good.' No one expected the miners to hold out for long, but that was because no one expected this strike to be different from others. At first the miners picketed coal-fired power stations, but then rapidly expanded their campaign to take in all power stations as well as other major users of coal, such as steelworks. There was general resentment to the government's pay freeze, so that many other workers were prepared to support the miners, including the dockers of South Wales who refused to unload coal that was being brought in from overseas. A new phenomenon appeared, the 'flying pickets': men who were prepared to move around the country to wherever they were needed. The miners of Tondu in Wales were able to go off because their wives took over local picketing duty; the picketing wives were able to stop clerks working in the NCB headquarters – it was another source of grievance that they were earning more than the men winning the coal.

The more moderate union officials found a new militancy among the rank and file. The moderate camp had been in favour of allowing safety men to carry on working. They were the men responsible for checking the pit in order to ensure that there was nothing liable to cause a roof collapse or flooding, essential if there was to be a pit to go back to once the strike was over. But the militant faction would have none of it, and stopped the safety men from going underground. There was a certain grim determination on both sides, which led to some ugly scenes and then tragedy. At Saltley coke works a truck driver tried to force his way through the picket line and a young miner, Fred Matthews, was knocked down and killed. Support for the strikers grew in the country and the conflict appealed to a new generation of left-wing students. Forbidden to use student union funds to contribute to the strike fund, they arranged 'lectures' by miners, for which they received unusually large fees. Miners picketing power stations in East Anglia were given accommodation on the University of Essex campus by the students, until the university authorities stepped in and turfed them out again.

The government were caught unprepared for the severity of the miners' actions and by 9 February 1972 they had declared a national emergency. Two days later they put the whole country on a three-day working week. It was not a situation that could be allowed to go on for long, and on 15 February they set up a committee under Lord Chief Justice Wilberforce to investigate the pay claim. It was probably the shortest inquiry of its time, for the results were with both parties within three days. On 19 February the two sides reached agreement, the pickets were called off and the pay offer put to the miners. It was accepted. The men got from £4.50 to £6 a week against the original top offer of £1.60. Apart from the fact that the strike resulted in a wage rise roughly treble what was originally on offer, it demonstrated that whatever the changes in the world were, coal was still a vital part of the economy. To the left wing it was a reassertion of the power of the rank-and-file workers. As one commentator put it, the strike 'illustrated clearly the weakness of the leadership and the great need for a strong rank-and-file membership'.[4] It was true that Joe Gormley was not particularly interested in playing politics – his concern was to get the best deal he could for the members, and it is worth noting how very rarely a miners' leader had actually achieved such a success. On the other hand, he was deeply distrustful of the more militant activists within the union. He was a compromiser by nature and very much on the right wing of the trade union movement. This was always well known, but what was only revealed as late as 2002 was that he had secretly been passing information about left-wing activists to the police Special Branch; an action that would have surely horrified even his most loyal supporters.

The results of the Wilberforce Committee gave the miners some of what they had hoped for, but it also cast a light on working conditions in the industry. There were still places where it seemed that, in many respects, things were no better than they had been a century before. One of those who gave evidence was a 42-year-old man who had worked in the industry for twenty-seven years, and was then at Snow Hill Colliery, Kent:

> I am working in a pit which is much hotter than the previous one. Indeed, the men at the pit where I now work wear no clothes at all when working. This was unusual to me when I went there because I was used to working in short trousers, but eight out of ten men in the headings work with absolutely no clothes on because of the heat and, because of the amount of sweating they do, they have to drink a lot of water. Many, many men at Snow Hill Colliery drink eight pints of water a day and the Coal Board has provided them with salt tablets to put in the water to stop them getting whatever they are supposed to get by drinking a lot of water … Since the advent of mechanisation in the industry the amount of dust which is in the places of work has to be seen to be believed. Dust-suppression methods are used but in many cases they are not effective. I work with a dust respirator to stop the dust from going into my lungs but a number of men do not wear these masks.

He travelled an hour each way to work, and started his shift at 4.30 a.m. He reported that he was actually earning 50p less per shift than he had been in 1963 and, as he said, that should help the committee understand the attitude of the pickets.

The 1972 settlement was seen by the unions as no more than an interim measure, and by early 1973 the miners had slipped down to eighteenth place in the industrial wage tables. The country was in a state of economic turmoil, with oil prices rocketing due to the Yom Kippur War between Israel and the Arab world. The government once again decided that the best way to reduce inflation was to freeze wages. When the miners were refused the pay rise that they felt was due to them, they imposed an overtime ban. Prime Minister Edward Heath responded by imposing a three-day working week and insisting that all television broadcasts had to stop early to save electricity. The theory was that, deprived of the television, people would go to bed earlier and find other entertainments, which a slight peak in the birth rate nine months later suggested that they did. He believed that the country as a whole supported his actions, so he called a General Election in the hope that it would give him a mandate to take far tougher action against the miners. Instead, he lost and Harold Wilson came into Downing Street to head a minority government. The new Secretary of State for Employment, Michael Foot, was soon able to reach an agreement and apart from making a decent pay award, he introduced a compensation fund for pneumoconiosis sufferers and a new superannuation scheme. It was a famous victory for the miners and a deeply humiliating defeat for the Conservatives; a humiliation that was neither forgotten nor forgiven. It was also a vindication of Joe Gormley's approach. Few trade union leaders in mining history had achieved as much. It was not a lesson everyone was prepared to learn.

CHAPTER SIXTEEN

THE LAST BATTLES

By the start of the 1980s, prospects were looking a little brighter for the beleaguered tin mines of Cornwall. The price for the metal was high and it seemed to be worth investing in opening up new, deep lodes. At Wheal Geevor, near the old Levant mine on the north Cornish coast, the company created an incline to deeper parts of the mine, and the improvements were given royal approval when the queen unveiled a plaque to mark the event on 28 November 1980. I visited the mine a couple of years later and there was a real spirit of optimism among everyone I met.

The start of the journey was exactly like that of any mine, a rapid drop down a shaft in the cage, but the first difference appeared at the bottom. Having been used to coal mines it came as a shock to find men smoking below ground, but there is no reason not to do so as far as mine safety is concerned. Tin mines do not have any dangerous gases waiting to ignite. The first impression was of an open area as brightly lit as a supermarket, but that soon changed as we left the shaft and began splashing away down passageways, ankle deep in orange, murky water. Like most mines, the pumps have to be kept going night and day, but there was a difference here: there are two pumps, one to remove fresh water, the other to take away salt water. It is a somewhat sobering thought that the sea is constantly trying to break in, to such an extent that they have to pump out 700,000L a day. That, however, is modest compared with the work of the fresh-water pump removing 1.6 million L a day.

Cornish mines had their own, to me, unfamiliar language rather different from that in use a century before. There were four different groups of men at work: rock breakers (no explanation needed); trammers and diggers; timber men; day pay men. The first working group I met were 'tramming a grizzly'. In other words, they were pushing a tram or small truck down a railed track and tipping the contents into a hole covered by widely spaced iron bars. The grill held back the bigger chunks which were then broken up with a hammer. The timber men were responsible for putting in props and constructing the staging, from which men attacked the ore. But I was still a long way from any sight of the metal itself, so there was more paddling through what looked increasingly like tomato soup. The strip lighting eventually came to an end, and now when I glanced up my helmet lamp spotlighted the thin line of shining ore in the roof of the passage – the lode the miners were following.

Finally, the echoing roar of drills made it clear that the face was near. I found two men, enveloped in clouds of dust, drilling into a solid wall of granite, getting ready for blasting. They moved this tunnel forward 2–3m every day, blasting away some 25 tons of granite in the process. These were the day pay men who, like generations of Cornish miners before them, were paid by results. The mine was a three-dimensional labyrinth: above, to the side and below other miners were similarly engaged, drilling the hard granite. I eventually went back to the surface up a different shaft, a far older shaft full of little bends – 'going up the banana' they call it. Back at the surface I stood on the cliffs looking out at the sea and the orange stain spreading out from the shoreline, the same ochre-stained water that I had been tramping through a short time before. I found it difficult to believe that it was not long ago that I had been walking far out under those waves. It had been an interesting experience and, talking to men and managers, I got the impression that they were looking forward to a prosperous future for mining in the area. They were to be bitterly disappointed.

In 1985, just months after my visit, the price of tin collapsed on the international market, from £10,000 a ton to £3,500. New, easily worked deposits had been discovered in Malaysia and Brazil and, at the same time, America had decided to put all its reserve tin stock on the market. Cornwall could never make a profit at the new prices and on 16 February 1990, Wheal Geevor closed down. The pumps were shut off and millions of gallons of water drowned out the mine. Today, parts of the mine are kept open as a museum, but the days of commercial mining in the south-west have come to an end.

In the coal industry, there was a certain sense of euphoria among the miners as a new 'Plan for Coal' was brought forward by the government, which seemed to promise better days ahead. Meanwhile, the Conservative Party was not sitting back bemoaning its defeat, but planning for future battles. A committee was set up under Nicholas Ridley to look at future policies. Its findings were reported in 1977.[1] Privatisation of all the nationalised industries was high on the list, which was to be accomplished 'by stealth'. The plan was to first pass legislation to break public sector monopolies and then to fragment the industries and sell them off, bit by bit. It was recognised that this would be fiercely resisted by the unions, and that the biggest threat would come from the miners. A strategy was set out to defeat the men who the report specifically referred to as 'the enemies'. The first objective was to fight on ground of the government's choosing, at a time that gave them maximum advantage. The tactic was to 'provoke a battle in a non-vulnerable industry, where we can win ... A victory on the ground of our choosing would discourage an attack on more vulnerable ground'. This would only be a preliminary to what was expected as the main bout: government *v.* the miners. The first priority was to build up coal stocks at vital sites, such as power stations, and to try to keep some ports like Felixstowe and Shoreham open for coal imports. In order to make sure stocks could be moved around the country, the government would make contact with non-union haulage companies prepared to drive through picket lines. On the pickets themselves, 'the only way to do this is to have a large, mobile squad of police who are equipped and prepared to uphold the law against the likes of the Saltley Coke-works mob'. The government would need to have a policy of preventing strikers and their families getting any sort of state benefits so that the whole cost of the strike should fall on union funds. For the time being, there was nothing to be done, but the plan was there ready and waiting for when the Conservatives got back into power.

There were fewer developments in the mining industry as the 1970s gave way to the 1980s. One very important change came when the union negotiated with the NCB over a new pay structure, where wages would reflect output. Joe Gormley was willing to accept the deal, but the membership as a whole voted against it. He then decreed that it was up to individual regions to decide whether or not they wanted to have this sort of agreement and, not surprisingly, the miners in the very productive Nottinghamshire coalfield voted in favour. The decision to overthrow the national union vote was challenged in the courts, but the court upheld it. It was a decision that was to have serious repercussions over the next few years. Gormley did, however, get agreement to accept a 9.3 per cent pay rise. Then in 1981 he retired. By now there was widespread resentment at what many ordinary miners saw as an executive who was overriding their democratic rights. When the leader of the Yorkshire union put his name forward on a very left-wing ticket, with a promise to listen to the rank and file, he received 70 per cent of the votes. His name was Arthur Scargill. He was a very different character from Gormley. Born into a mining family – his father was also an active communist – he went down the pit as a young boy and joined the Young Communists. He later joined the Labour Party but never lost his left-wing convictions, nor his dislike of compromise. Another election two years earlier had brought another new leader to the country. Margaret Thatcher was like Scargill in just one respect: she was convinced of the rightness of her own views and was ready to fight for them. Situated at either end of the political spectrum, the clash of personalities and creeds was bound to result in conflict, sooner or later.

In 1983 Ian MacGregor was appointed as the new head of NCB. He was a man who had earned his reputation by turning round the fortunes of British Steel Corporation, but only at the expense of making nearly half the workforce redundant. He was expected to show similar ruthlessness in his new job. The NUM responded by imposing an overtime ban in October that year. Ostensibly, the main objective was to cut production to preserve jobs, but it was also aimed at reducing coal stocks. The government had no intention of doing anything immediately; as the Ridley Report had suggested, the best time to fight the miners was the spring when, provided there were ample coal stocks, the demand would be at its lowest. Ignoring agreed procedures for consultation with the NUM, the National Coal Board announced the closure of Cortonwood Pit to take effect from 1 March 1984. There was no obvious, rational reason to select this particular colliery. The men had been told their jobs would be secure for at least another five years and a lot of money had just been spent on refurbishing the pithead baths. It did, however, have one obvious appeal to Thatcher and MacGregor: the pit was in South Yorkshire, at the very heart of Scargill's home territory. They were throwing down the gauntlet: if he failed to rise to the challenge they could carry on with a far wider programme of closures with some confidence. If he decided to fight, then there could not be a better time for the government to confront the man who Thatcher was calling 'the enemy within'.

Arthur Scargill called for a national strike, but followed the example of Joe Gormley by declaring that it was not necessary to hold a national ballot. It turned out to be a major error of judgement. Many of the Nottingham miners decided that, as they had not been allowed to vote in a national poll, they had no need to follow the executive's decision. They worked on. The Labour Party and several other trade unionists, who felt that the strike was ill timed, also used the issue of the national ballot as an excuse to offer only the feeblest support to the miners. It was not the first time the miners had acted alone, but this time there was disunity within their own ranks. At first, it seemed that the strike might follow the pattern of 1972, but the government had their new tactics ready. Apart from the fact that coal supplies had not been effectively cut off, they had a new weapon to deploy: the riot police.

Busloads of pickets were stopped on main roads and turned back at police roadblocks. The miners did reach an agreement with British Steel and the local unions to allow some coal through to their coking plants – coke was the essential fuel needed to keep the furnaces running. Then they discovered that far more coal was being delivered than had been agreed, so the miners' leaders decided on a mass picket at Orgreave coking plant in South Yorkshire. What happened next went down in mining history as the Battle of Orgreave. There are two versions of what happened on 18 June 1984: one supplied by the police and the government, and the other by the miners and their supporters. The official view was clearly expressed when several of the strikers were brought to court, and charged with various forms of violence and rioting. The prosecution began their case with a statement about why the men had come to Orgreave: 'Their aim was force and violence. No miner on the 18th could be ignorant of the situation. He would have to have been as naive as a babe in arms.' One of the defendants, Michael Wilson, gave his version of what happened to him that day:

I was standing here talking to my mate about going to see a pigeon fancier on the allotments over the way, and at that particular moment I see the ranks open up and there's about six policemen coming out with riot shields. They dropped the riot shields on the road and started to run towards me. As they got about five or six yards off me I realised they were going to hit and so I turned and ran. And I went down here and as I went down, there was thump, thumping sound on the temple. I could feel blood running down on the side of my temple. I could feel blood running down here … Eventually they got me up after regaining consciousness and walked me through the police lines.[2]

A very different view of events was given by Assistant Chief Constable Anthony Clements in evidence. He claimed that 'Orgreave was policed in a humane and orderly manner' and that police never hit anyone who was running away, but only in self-defence. They were, he said, under attack: 'It's no exaggeration to say that the sky was black with missiles. Bottles, heavy machinery, ball bearings. So I sent the horses in again.' It is difficult to believe that they are describing the same events on the same day. So what did happen at Orgreave?

Up until then, the police had prevented miners from outside the area getting anywhere close to the site to be picketed, but on this occasion they were ushered into a field, where thousands gathered. When the first lorries arrived, the pickets began to push forward and mounted police were sent in. The miners fell back, and the horses retreated, leaving a 30yd gap, through which the trucks made their way to the plant. The process of pushing the pickets back was repeated, but this time the line of police moved forward as well. Then the pickets were ordered to retreat again a full 100yd, and this time the police armed themselves with full riot gear, short shields and batons. The police charged again. It seemed as if things had quietened down by the end of the morning, as many of the miners headed off to the village for refreshments. The rest were lounging around in the field when the police charged once more, later claiming they were provoked by violence from the pickets. There were more charges, some fighting and the pickets were driven off the field, down a railway cutting. Arthur Scargill fell down the bank, either hit by a riot shield according to the miners' account, or by slipping according to the police. Eventually, the battle was continued out into the village, where the pattern of charges by police and mounted police continued until the miners were routed. Some accounts say that the miners were pursued and beaten as they ran away.

The events were filmed and photographed. The first charge by men in riot gear was supposed to have taken place when the air was black with missiles; if so they remained strangely invisible in film and photographs. The claim that the police only attacked with batons in self-defence is greatly weakened by a famous photograph, showing one of the Women Against Pit Closures, Lesley Boulton, standing with a camera in her hand while a mounted policeman bears down on her, leaning out of the saddle, baton raised to strike. A second photograph shows her being pulled away as the baton just misses her. She later said that there was no provocation for the afternoon attack: 'A lot of the men had taken their t-shirts off and stuffed them in their back pockets. It certainly wasn't the sort of thing you'd do if you were planning to attack a seriously armed police force.' There seems very little doubt that, although there was some retaliation by the miners in terms of stone throwing, the most violent attacks were carried out by officers of the law themselves. Many believed that the truth would finally emerge when the men charged with rioting were brought to court.

Ninety-five pickets were charged with riot and unlawful assembly. The court hearing never got very far, however. Defence council, Michael Mansfield QC, began questioning the police evidence. The accused, many of whom had received head injuries and one of whom had a broken leg, described how they were bundled back behind the police lines where two detectives dictated what the constables should put in their notebooks. In the case of one miner, a report had been signed by two officers and the defence claimed one of the signatures was a forgery. It then mysteriously went missing, but copies were made available and an expert witness confirmed the forgery. The judge ordered that the miner be acquitted. More and more of the police evidence was investigated and failed to stand up to scrutiny. In the end every case was thrown out, before the defence even got round to making their case. While the accused were no doubt relieved not to be found guilty, they were disappointed not to be able to give their own version of events. There was to be a further vindication of the pickets' version of events. In an out-of-court settlement, South Yorkshire Police paid out £425,000 to thirty-nine pickets as compensation for unlawful arrest. It seemed to be the most telling indicator of what really happened, but events elsewhere provided fresh grounds for doubting the police version of events.

The whole question of who was telling the truth about Orgreave was raised again in 2012, following the report by an independent panel on the Hillsborough disaster of 1989, in which ninety-six football supporters from Liverpool lost their lives. It emerged that there had been a deliberate attempt to blame the fans for what happened and blacken their reputations. The panel found that 164 police statements had been altered. This was the same South Yorkshire police force under the same chief constable that had been involved at Orgreave. The matter has now been referred to the Independent Police Complaints Commission, and the NUM and Michael Mansfield QC have demanded that it should also be investigated by the Director of Public Prosecutions.

Back in 1984, the weeks of the strike lengthened into months and it became clear that the strategy outlined in the Ridley Report was proving effective. There were no major disruptions to power supplies, partly thanks to stockpiling and also because the majority of Nottinghamshire pits, among the most productive in the country, remained open. The Nottingham union leaders were incensed by Scargill's refusal to hold a national ballot and eventually they were to break with the NUM to form the Union of Democratic Mineworkers. The striking miners suffered a great deal as they tried to live on just their strike pay, as recent government legislation had stopped not just striking miners but also their dependants from getting any money from public funds. In spite of the hardships, the families remained united, and miners' wives played an active part in supporting the strike, even if their kitchen cupboards were looking ever emptier. But there was help from around the country. Many local Labour Party members got together to collect food to take to the mining communities. The Oxfordshire town where I lived at the time was one of those, and weekly deliveries were made to South Wales. It was obvious that non-perishable food would be best, but a message did eventually come back that while everyone was grateful for the help, would it perhaps be possible to have rather fewer tins of baked beans. At the opposite extreme, a neighbour had forgotten that it was collection day, and handed over the only tin he had in the house – baby octopus. No one ever discovered whether it was snapped up by a mining gourmet or was left unopened on a shelf.

As time passed and with the possibility of final victory for the miners looking less and less likely, there was a slow drift back to work. Inevitably, there was a great deal of bitterness among the men who stayed out. It led to one of the most shameful acts of the whole dispute. David Wilkie was a taxi driver who had agreed to take a miner, with police escort, from his home in Rhymney to a colliery at Merthyr. In November 1984, two miners pushed a lump of concrete off a bridge over the A465 on to the cab. Wilkie was killed, and the two men were found guilty of murder, later reduced to manslaughter. It was not the only death: six pickets died and three teenagers. Towards the end of the strike, as funds got ever lower, real poverty overwhelmed many families and in an effort to try and keep warm, several tried scavenging for bits of coal from the heaps that overshadowed so many collieries. These were notoriously unstable, as had been tragically demonstrated at Aberfan on 21 October 1966 when the tip collapsed, sliding down to engulf first a cottage then the Pentglas Junior School. Altogether 144 were killed that day, including 116 school children. The dangers of scavenging were well known, but these were desperate days: the three teenagers were engulfed when the tip collapsed on top of them.

By early 1985 it was clear that funds were no longer sufficient to support the strike: families were going hungry; and there was not enough cash to send out buses for pickets to keep up what pressure they could. On 3 March the strike was officially ended. The men went back to work, marching to the collieries with their banners flying and the bands playing. It was a brave display, but nothing could disguise the fact that it was an utter and complete defeat. They had been out for almost exactly a year and they had not even been able to negotiate a single concession. There was nothing at all to sugar the very bitter pill they had been forced to swallow. Inevitably, there were many who looked back on the events and asked how things had gone so wrong. Many saw one of the key factors as being Arthur Scargill's failure to hold a national ballot. Whatever the legalities of the situation, there is no doubt that it was the key factor that kept the Nottinghamshire men, who had demanded the ballot, at work. The timing of the start of the strike could scarcely have been worse: successful coal strikes in the past had all started with the approach to winter when demand was at its peak. To some extent the start had been forced on the NUM by the government's actions in closing Cortonwood Pit, but a more cautious leadership might have seen this as the trap laid to ensnare them.

In 2011 the UK consumed 51.2 million tons of coal, of which 32.4 million tons was imported, mainly from Russia. The majority of the coal was used in power stations.

Following the collapse of the strike, the closures began, with twenty-five pits being shut that year. It was just the start and a preliminary to the government's long-term aim of privatising the industry. The Coal Industry Act of 1994 made this possible, transferring all the remaining collieries to a new organisation, Central and Northern Mining Ltd, now UK Coal. This did nothing to halt the closures. Scotland lost all its collieries, and South Wales looked like going the same way when the closure of the last remaining pit, Tower Colliery, was announced in April 1994. But it was not to be the end for Tower Colliery. The 239 miners who were losing their jobs put in their redundancy money, raising over £1 million to keep the pit open. It prospered for a time, but after a decade of work the profitable seams began to peter out. It was finally closed on 25 January 2008. Today only three deep pits remain open in England, producing 1.7 million tons of coal a year, another 5.8 million tons coming from opencast, now known as surface, workings. How incredible these figures would have seemed a century ago. H. Stanley Jevons, writing in 1915, included a table of projected output for Britain's coalfields in his book *The Coal Trade*. His estimate for 2011 was 784 million tons. He could never have imagined that production would be a mere 10 per cent of what he predicted, and certainly not that we would actually be importing much more coal than we produced, much of it coming from as far away as Australia.

Mining for coal and minerals had been at the heart of the British economy for centuries; today, the industry has been pushed to the outer margins. Should one mourn its passing? I often drive up the M1 motorway, and along the way I pass an artificial hill. This grassy mound was once a colliery tip, all that remains of Markham Main Colliery. However, my thoughts are never really about the men who lost their jobs when the pit closed – I vividly remember hearing the news in 1973 that the brakes on a cage had failed, plunging eighteen men to their death. It was not the first, or even the worst, disaster at this colliery. In 1938 seventy-nine had been killed and forty injured in an explosion. No one can be pleased at the fact that thousands of men have been thrown out of work, that communities have lost their focus, their reasons for existing. No one who has ever spent any amount of time with miners could fail to be impressed by their pride in doing a job that most of us would never contemplate. But there is something else that goes with that pride; time and again you would hear a miner say that no matter what he thought about his work, he did not want his son to follow in his footsteps underground. The prosperity of the past was purchased at a high price; it was paid for in sudden death from accident and lingering, painful death from lung and heart disease. No one will regret if those days are gone forever, as far as this country is concerned. However, someone, somewhere is still toiling underground, producing the coal and the metals, and all too often in conditions that are every bit as bad as those endured by the men, women and children of the British mining community two centuries and more ago. Sadly, the story of oppression in mines has not ended; it has simply moved somewhere else.

The lives of the miners were largely passed in a world that others knew little about, and this was something that could make a man bitter. These are the words of a miner who had just emerged from a visit to the doctor, where he had learned that ill health was about to bring his working life to a close:

An hour later I was sitting in the Victoria Gardens at Neath. This is a lovely little park just away from the busy streets. The flowers delight the eye and above it all the clock in the tower of St. David's Church warned us of the passing of each bright hour, and the sun was genial that day. Out in the warmth the people moved about their lives. The shops were busy, and loaded lorries whined past. This was normal life to them, the only way they understood. What did they know of roof falls, or stiff working coal, or men who gasped as they breathed? What did they care while the sun shone and the world was bright? Yet nearly three-quarters of a million men – my mates – were shut away from that sun. Their normal life was about them then, in a crushing darkness with sweat running down their backs to make their singlets like a wet cloth. Men straining to rip out the compressed sunlight which had been stored in the heart of creation uncounted centuries before; and which to a greater or a lesser degree affects all our lives and our national prosperity.[3]

Mining may no longer have that same direct effect on all our lives that it did when those words were written more than half a century ago, but we should never forget that the industrial world was built on the efforts of generations of workers, labouring in dark places far beneath the ground.

NOTES

ONE – Beginnings

1   A full account can be found in C. Andrew Lewis' MPhil thesis,
    University of Wales, Bangor, 1956.
2   Quoted in Frank Nixon, *Industrial Archaeology of Derbyshire*, 1969.
3   A full account can be found in P.R. Lewis, 'The Ogofau Roman Gold
    Mines at Dolaucothi', *The National Trust Year Book*, 1976–77.
4   Rowlan Price, *National Association of Colliery Overmen, Deputies &
    Shotfirers, Midland Area: A Short History*, 1962. Unless otherwise stated the
    following quotations are all taken from these documents.

TWO – T'owd Man

1   Agricola, *De Re Metallica*, 1556.
2   Quoted in H.C. and L.H. Hoover, *The Mining Magazine*, 1912.
3   Quoted in Frank Nixon, *Industrial Archaeology of Derbyshire*, 1969.
4   Frank Booker, *Industrial Archaeology of the Tamar Valley*, 1967.
5   Quoted in A.K. Hamilton Jenkin, *The Cornish Miner*, 1927.
6   Richard Carew, *Survey of Cornwall*, 1602.
7   John Holland, *Fossil Fuel*, 1835.
8   Robert L. Galloway, *A History of Coal Mining in Great Britain*, 1882.
9   Quoted in William Pryce, *Mineralogia Cornubensia*, 1778.

THREE – The Deep Pits

1   William Craig, *Chorographica*, 1649.
2   Daniel Defoe, *A Tour through the Whole Island of Great Britain*, 1724–26.
3   Robert Bald, *General View of the Coal-Trade of Scotland*, 1812.
4   Thomas Wilson, *Pitman's Pay*, 1843.
5   Holland, *op. cit.*
6   Quoted in J.C., *The Compleat Collier*, 1708.

FOUR – Birth of the Steam Age

1   *Post Man*, 19–21 March 1702.
2   Quoted in article by David Tyler, *Transactions of the Newcomen Society*, Vol. 76, No. 1, 2006.
3   Matthias Dunn, *An Historical, Geological and Descriptive View of the Coal Trade of the North of England*, 1844.
4   John Hodgson, *An Account of the explosion at Felling*, 1813.
5   Quoted in Anthony Burton, *The Rainhill Story*, 1980.
6   Roy Palmer (ed.), *Poverty Knock*, 1974.

FIVE – Masters and Men

1   John Taylor, 'On the Economy of the Mines of Cornwall and Devon', *Transactions of the Geological Society*, 1814.
2   L.L. Price, *West Barbary*, 1891.
3   Quoted in Jenkin, *op. cit.*
4   Sidney Webb, *The Story of the Durham Miners*, 1921.
5   Dunn, *op. cit.*
6   Quoted in J.L. and Barbara Hammond, *The Town Labourer*, 1917.

SIX – The First Unions

1   Parliamentary Papers: Combination Act, 1800, 1839 and 1840 Geo. III, *c.* 106.
2   Quoted in Anthony Burton, *The Miners*, 1976.
3   *Annual Register*, 1817.
4   *Address to the Colliers of Ayrshire*, 1824.
5   *Report from the Select Committee on Combination Laws*, 16 June 1825.
6   Quoted in David Bremner, *The Industries of Scotland*, 1869.
7   Sidney and Beatrice Webb, *The History of Trade Unionism*, 1920.
8   A.L. Lloyd (compiler), *Coaldust Ballads*, 1952.

SEVEN – Women and Children

1   Rev. John Hodgson, *Picture of Newcastle upon Tyne*, 1807.
2   Collins, *op. cit.*
3   *The Political Register*, 20 July 1833.
4   Royal Commission Reports on Children in the Mines, 1842.
5   Georgina Battiscombe, *Shaftesbury*, 1975.

EIGHT – Searching for Safety

1   Robert L. Galloway, *Annals of Coal Mining and the Coal Trade*, Series 2, 1836–50.
2   *Ibid.*
3   Robert L. Galloway, *A History of Coal Mining in Great Britain*, 1882.
4   *British Miner*, December 1862.
5   Charles Williams, *Buried Alive!*, 1877.

NINE – The Fight for Unity

1   Richard Fynes, *The Miners of Northumberland and Durham*, 1873. The subsequent quotes and description of the 1844 strike are from the same source.
2   Quoted in O.P. Edmunds and E.L. Edmunds, *British Journal of Industrial Medicine*, 1963.
3   Letter to *The Flint Glass Makers' Magazine*, October 1851.
4   Frederick Engels, *The Condition of the Working Class in England*, 1892.
5   Quoted in Raymond Challiner, 'Alexander MacDonald and the miners', *Our History*, No. 48, 1967.
6   Alexander Dalziel, *The Colliers Strike in South Wales*, 1872.

TEN – Above and Below Ground

1   Anon, *Life Among the Colliers*, 1862.
2   Quoted in Jenkin, *op. cit.*
3   R. Quiller Couch, *Statistical Investigation of the Mortality of Miners*, 1856.
4   Jenkin, *op. cit.*

ELEVEN – Union

1   Quoted in R.P. Arnot, *The Miners*, 1949.
2   A full account can be found in Frank Booker, *Industrial Archaeology of the Tamar Valley*, 1967.

TWELVE – Two Wars

1   The account was written by Boothby for his old school's magazine and is quoted in Arthur Stockwin (ed.), *Thirty-odd Feet Below Belgium*, 2005.
2   Minutes of the meeting of the War Cabinet, 27 September 1917.
3   H. Stanley Jevons, *The Coal Trade*, 1915.
4   William Brace *London Magazine*, October 1914.

THIRTEEN – A Time of Struggle

1   Quoted in Christopher Farman, *The General Strike*, 1972.
2   *Ibid.*
3   Quoted in William Camp, *The Glittering Prize*, 1960.
4   Jarvis, *op. cit.*
5   J.B. Priestley, *English Journey*, 1934.

FOURTEEN – Depression and War

1   George Orwell, *The Road to Wigan Pier*, 1937.
2   Orwell, *op. cit.*
3   www.museumwales.ac.uk/en/rhagor.
4   A.K. Hamilton Jenkin, Introduction to the 3rd edition of *The Cornish Miner*, 1962.

FIFTEEN – Coal for the Nation

1   Minutes of Cabinet meeting, 13 December 1945.
2   Quoted in Michael Foot, *Aneurin Bevan*, 1975.
3   Memorandum to the Cabinet, 27 December 1945.
4   John Charlton, *International Socialist*, April 1973.

SIXTEEN – The Last Battles

1   Report of Nationalised Industries Policy Group, 1977.
2   Interviewed in *The Battle of Orgreave* documentary film, Journeyman.tv.
3   B.L. Coombes, *Miners Day*, 1945.

INDEX